Math Mammoth Grade 7 Skills Review Workbook

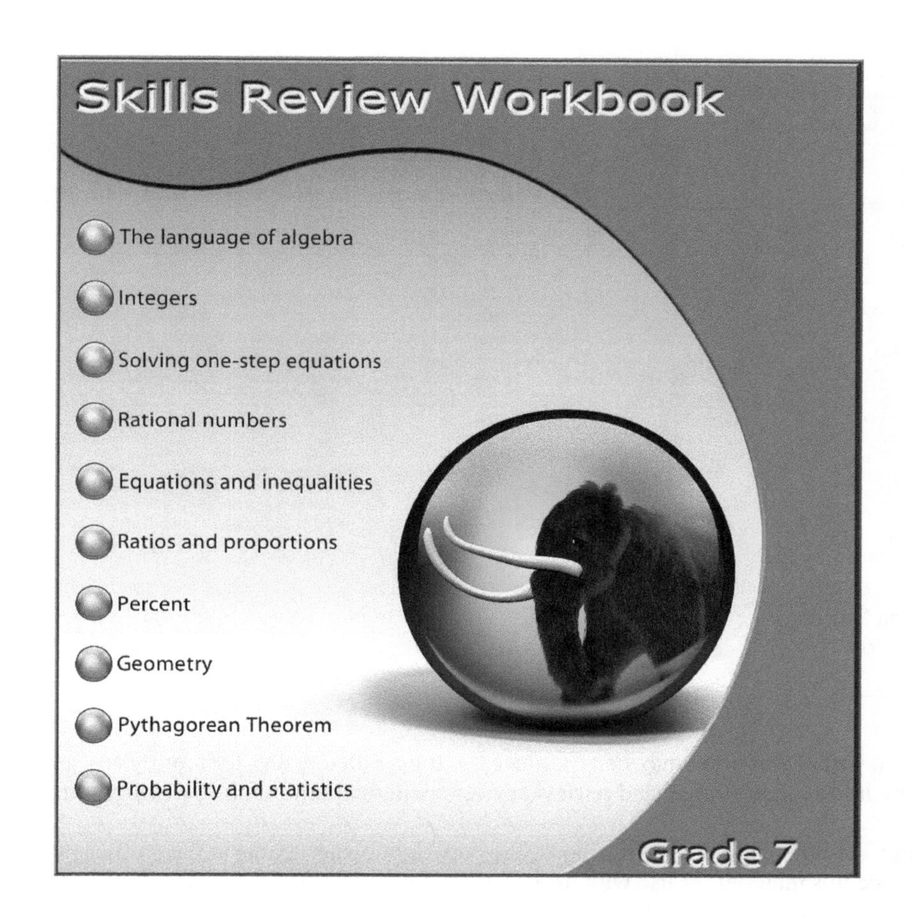

By Maria Miller

ISBN 978-1-942715-76-4

Edition 8/2020

Contents

Chapter 4: Rational Numbers

Chapter 5: Equations and Inequalities

Chapter 6: Ratios and Proportions

Chapter 7: Percent

Chapter 8: Geometry

Chapter 9: Pythagorean Theorem

Chapter 10: Probability

Chapter 11: Statistics

Foreword

Math Mammoth Grade 7 Skills Review Workbook has been created to complement the lessons in *the Math Mammoth Grade 7* complete curriculum. It gives the students practice in reviewing what they have already studied, so the concepts and skills will become more established in their memory.

These review worksheets are designed to provide a spiral review of the concepts in the curriculum. This means that after a concept or skill has been studied in the main curriculum, it is then reviewed repeatedly over time in several different worksheets of this book.

This book is divided into chapters, according to the corresponding chapters in the *Math Mammoth Grade 7* curriculum. You can choose exactly when to use the worksheets within the chapter, and how many of them to use. Not all students need all of these worksheets to help them keep their math skills fresh, so please vary the amount of worksheets you assign your student(s) according to their needs.

Each worksheet is designed to be one page, and includes a variety of exercises in a fun way without becoming too long and tedious.

The answer key is available as a separate book.

I wish you success in teaching math!

Maria Miller, the author

Skills Review 1

1. Evaluate.

a. 0.3^3 **b.** 2^5 **c.** 0.1^4

2. Find the value of these expressions.

a. $\dfrac{1}{12-6} \cdot \dfrac{20}{40}$	**b.** $\dfrac{7 \cdot 8}{2} - \dfrac{2 \cdot 9}{3 \cdot 2}$	**c.** $1 - \dfrac{7-5}{4+2}$

3. Rewrite each expression using a fraction line, then simplify. Compare the expression in the top row with the one below it. *Hint: Only what comes right after the "÷" sign goes into the denominator.*

a. $3 \cdot 6 \div 2$	**b.** $2 + 24 \div (3 + 1)$	**c.** $42 \div 6 + 1 \div 5$
d. $3 \div 6 \cdot 2$	**e.** $2 + 24 \div 3 + 1$	**f.** $12 - 1 + 2 \div 3$

4. Write an expression.

a. y minus x squared.

b. The quantity y minus x, squared.

c. The quotient of x minus 5 and x to the third power.

5. **a.** A wooden board is m units long, and another board is 3/4 as long.
Write an expression for the length of the second board.

b. Write another, different expression for the length of the second board.
(Hint: if you used a fraction in a., use a decimal now, or vice versa.)

c. Write an expression for the total length of the two boards, if put end-to-end.
Then simplify it.

Skills Review 2

1. Circle the equation(s) that matches the situation. You do *not* have to solve the equation.

   The amount of land used for parks decreased by 1/5, and now it is 2,600 acres.

$\frac{5p}{4} = 2{,}600$	$\frac{4p}{5} = 2{,}600$	$p - 1/5 = 2{,}600$	$\frac{p}{5} = 2{,}600$	$p - (1/5)p = 2{,}600$

2. Find the root(s) of the equation $x^2 - x = 12$ in the set of positive even numbers that are less than 10.

3. Are the expressions equal, no matter what value n has? Give n some test values to check.

a. $\frac{n}{2} + \frac{n}{2}$ n	**b.** $(10 - n) - 1$ $10 - (n - 1)$	**c.** $(10 - n) + 1$ $10 - (n + 1)$

4. Write an equation. Then solve it.

   a. The product of a number and 5 is 75.

   b. One decreased by a certain number is two sevenths.

5. Find the value of these expressions.

a. $(0.5 + 0.5) \cdot (9 - 7)^4$	**b.** $(0.7 - 0.3) \cdot 4^2$	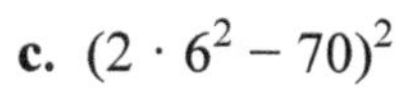 **c.** $(2 \cdot 6^2 - 70)^2$

6. Write a single mathematical expression ("number sentence") for each situation. Don't write just the answer.

a. The side of a square is $2x$ units long. Write an expression for its perimeter. *perimeter* =	**b.** Cindy earns \$15 per hour, and \$20 per hour for overtime. Write an expression for how much she earns for working x hours with the normal pay and y hours overtime. *earnings* =

Skills Review 3

1. Write an expression.

   a. the sum of a squared and b squared

   b. the quantity s minus 2, squared, minus 10

   c. 10 less than the quantity x plus 3

2. Which expression from the right matches with (a) and (b) below?

   a. The area of a square with sides 3 cm long.

   b. The volume of a cube with sides 3 cm long.

(i) 27 cm^3	**(ii)** (3 cm)^2	**(iii)** 9 cm^3	**(iv)** 3 cm^2

3. Simplify the expressions.

a. $a + a + 7 + n + 2n$	**b.** $5 \cdot y \cdot 4 \cdot y \cdot ½$	**c.** $3x \cdot x \cdot y \cdot y \cdot x \cdot x$

4. Evaluate the expressions. (Give your answer as a fraction or mixed number, not as a decimal.)

a. $5x - x^2$, when $x = 1$	**b.** $3mx - 2 + 2m$, when $m = 10$ and $x = 5$
c. $\dfrac{y^2 - 1}{y - 1}$, when $y = 3$	**d.** $\dfrac{z}{z - 1} + 1$, when $z = 5$

5. Find the value of these expressions.

a. $2 \cdot (4 - 2)^2 \cdot (200 - 100)^2$	**b.** $9 + \dfrac{2 \cdot (5 - 3)^3}{2^3}$	**c.** $\dfrac{(1 + 1)^4}{4} + \dfrac{15 - 12}{2}$

6. Some of these are wrong. Find which ones, and correct them.

a. $r \cdot r \cdot r = 3r$	**b.** $7 \cdot x + x = 7x^2$	**c.** $\dfrac{x}{x} = 1$

Skills Review 4

1. Which expression from the right matches with (a) and (b) below?

   a. The area of a square with sides $4x$ long.

   b. The volume of a cube with sides $4x$ long.

   (i) $4x^3$ **(ii)** $(4x)^2$ **(iii)** $(4x)^3$ **(iv)** $4x^2$

2. Evaluate the expression for (a) above when x has the value of 0.5 miles.

3. Camera batteries are discounted by \$5 from their original price. Mark buys four of them. Write an expression for the total cost he pays, if the normal price for one battery is p.

4. Are the expressions equal, no matter what values n and m have? If so, you don't need to do anything else. If not, provide a counterexample: specific values of n and m that show the expressions do NOT have the same value.

a. $n \cdot m \cdot 12$ $2 \cdot n \cdot 6 \cdot m$	**b.** $(n + 10) - 5$ $5 - (10 + n)$

5. **a.** Draw the next steps for the pattern below.

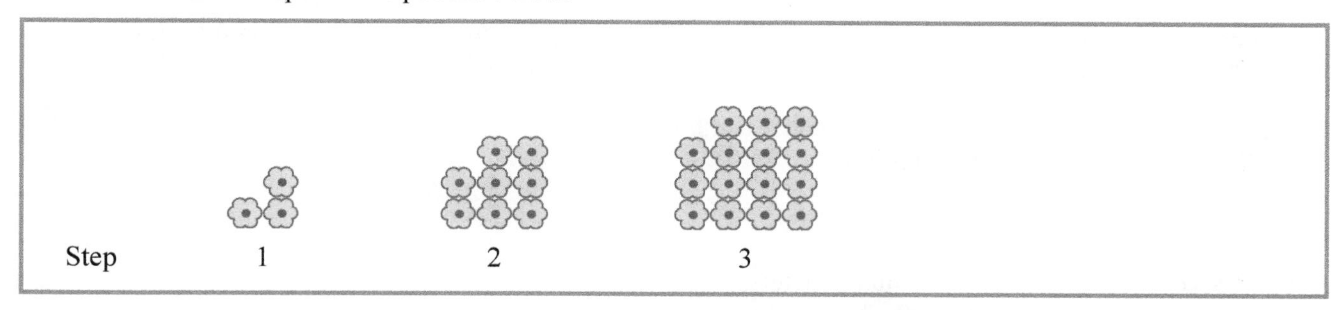

   b. How do you see this pattern growing? (There is more than one way to look at it!)

   c. How many flowers will there be in step 39?

   d. What about in step n?

Skills Review 5

1. Find the value of these expressions.

a. $60 \cdot (5-2)^2 - 20^2$	b. $11 + \dfrac{65-56}{2 \cdot (7-4)^3}$	c. $\dfrac{10^2}{(2+3)^2} \cdot \dfrac{23-8}{3}$

2. Use the distributive property to multiply.

a. $7(a - 11) =$	b. $200(t - s - ½) =$	c. $0.4(x + 2y + 7) =$

3. The perimeter of a square is $52m + 8$. How long is its side?

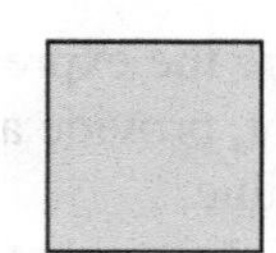

$P = 52m + 8$

4. Fill in the table.

Expression	the terms in it	coefficient(s)	Constants
$2x^5$			
$7x + 14y + 8$			
$a + 0.97$			

In the next problem, write an expression for part (a), and then for part (b) write an equation and solve it. You don't have to use algebra to solve the equation—you can solve it in your head or by guessing and checking.

5. **a.** The length of a rectangle is 4 meters and its width is w.
What is its area? Write an expression.

b. Let's say the area has to be 76 square meters.
How wide is the rectangle then?
Write an *equation* for this situation, using your expression from (a), and solve it.

6. Use the distributive property "backwards" to write the expression as a product. This is called **factoring**.

a. $14x + 49$ = ____ $(2x + 7)$	b. $54y - 12$ = 6(____ − ____)
c. $36y + 45$ = ____ $(4y +$ ____)	d. $108d + 60$ = ____ $(9d +$ ____)

Skills Review 6

1. **a.** A website offers both monthly and yearly subscriptions. The cost of a one-year subscription is \$50 less than 12 times the cost of a one-month subscription. If the cost of a one-month subscription is m, write an <u>expression</u> for the cost of a one-year subscription.

   b. Which of the equations lets you calculate the cost of a one-month subscription (m) if the cost of the one-year subscription is y?

   Hint: Give y *some test values.*

   $m = y + 50/12$ $\qquad$ $m = y/12 + 50$ $\qquad$ $m = (y + 50)/12$

2. **a.** Name the property of arithmetic illustrated by the fact that $(n + m) + 5$ is equal to $n + (m + 5)$

   b. Name the property of arithmetic illustrated by the fact that $t(2 + s)$ is equal to $(2 + s)t$.

3. Simplify.

   a. $|3|$ $\qquad$ **b.** $-(-11)$ $\qquad$ **c.** $-|-4|$ $\qquad$ **d.** $-|9|$ $\qquad$ **e.** $-(-(-3))$

4. **a.** Circle the equation that matches the situation. *Hint: give the variable(s) some value(s) to test the situation.*

   In a computer game, you gain 100 points every time you reach a new level. Also, 6 points are deducted for every error you made in that level. After finishing level 1, and making x errors, your total point count comes to 46.

$46 = 100 - x$	$46 = 6x - 100$
$100 - 6x = 46$	$x(100 - 6) = 46$

   b. How many errors did you make?

5. Plot these inequalities on the number line.

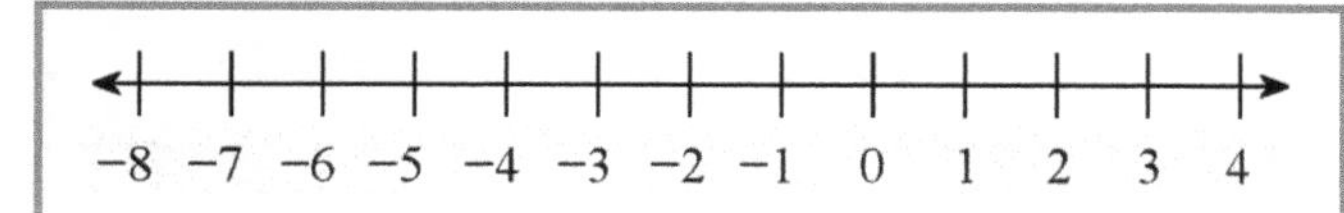

a. $x > -2$

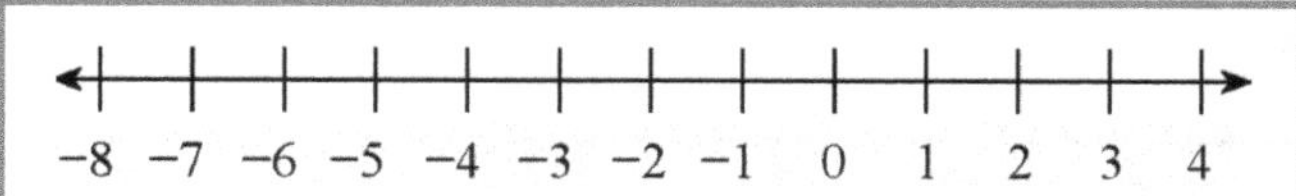

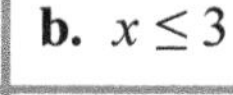

b. $x \leq 3$

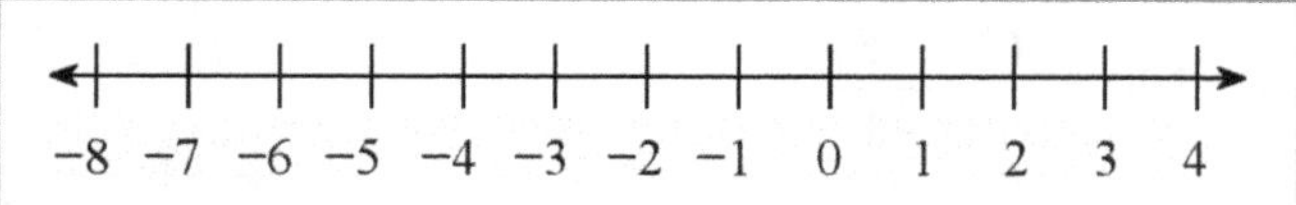

c. $x < -1$

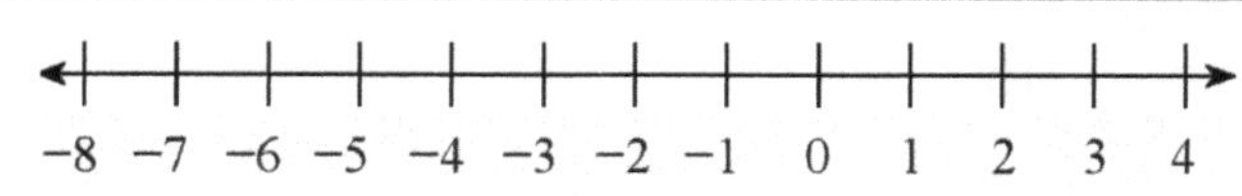

d. $x \geq -5$

Skills Review 7

1. Draw a number line jump for each addition or subtraction, and find the answers.

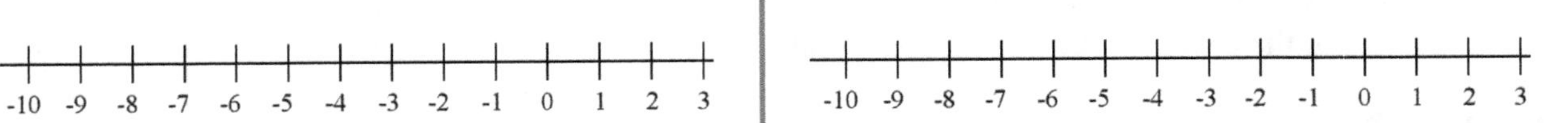

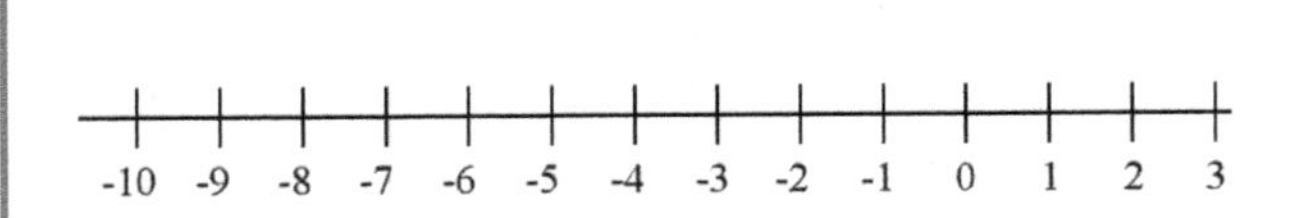

a. $-2 - 7 =$ ______

b. $-4 + 7 =$ ______

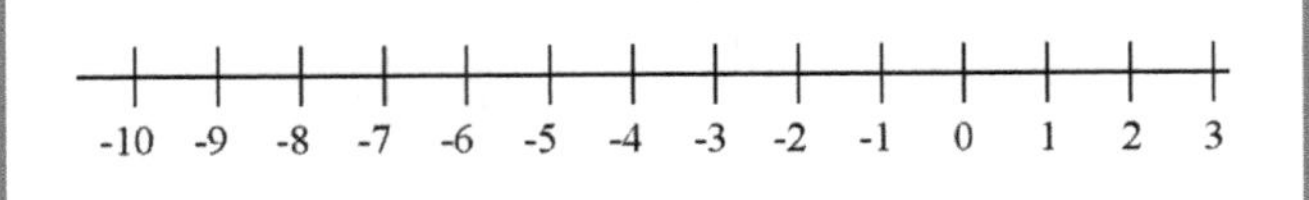

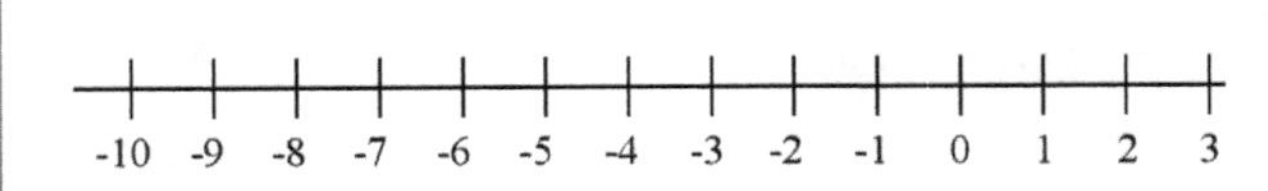

c. $-3 + (-2) =$ ______

d. $1 + (-6) =$ ______

2. Write an addition or a subtraction.

Addition/subtraction:

a. You are at $^{-}5$. You jump 9 to the right. You end up at ______.

b. You are at $^{-}1$. You jump 10 to the left. You end up at ______.

3. **a.** Sketch a rectangle with sides $2m$ and $5m$ long.

b. What is its perimeter?

c. Divide your rectangle in half, so that you get two equal areas.
What is the area of one of those halves?

4. Write an expression for the area using the distributive property, and then simplify it.

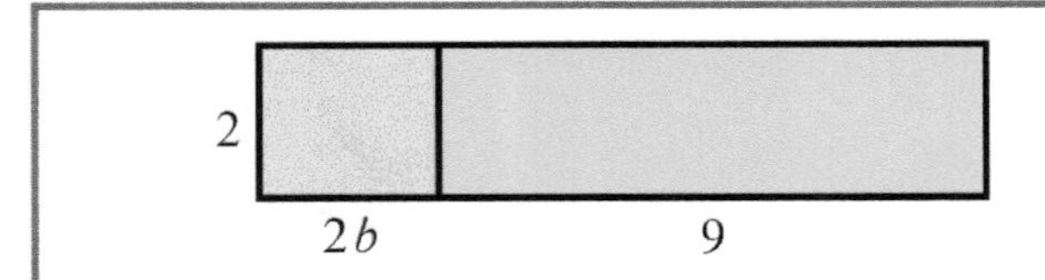

a. ____(_____ + _____) =

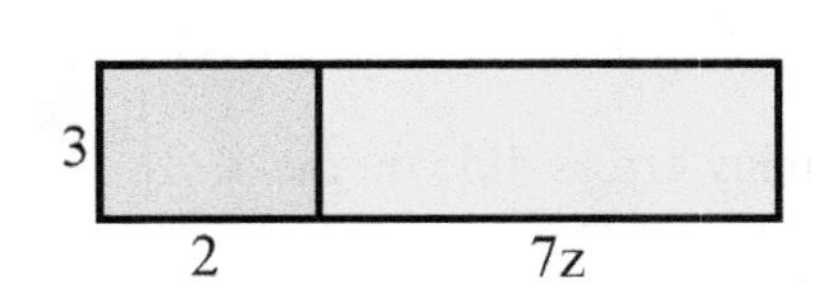

b. ____(_____ + _____) =

5. Find the value of these expressions.

a. $\frac{3^4}{3^2} \cdot (100 - 90)$

b. $100 - \frac{(12 - 5)^2}{8 - 1}$

Skills Review 8

1. Add. You can think of counters or number line jumps.

a. $-7 + (-7) =$	**b.** $3 + (-6) =$	**c.** $50 + (-13) =$	**d.** $-13 + (-14) =$
$18 + (-7) =$	$-8 + 5 =$	$-40 + 14 =$	$10 + (-12) =$

2. Compare how $-4 + 7$ is modeled on the number line and with counters.

   a. On the number line, $-4 + 7$ is like starting at _____, and moving _____ steps to the ____________, ending at _____.

   b. With counters, $-4 + 7$ is like _____ negatives and _____ positives added together. We can form _____ negative-positive pairs that cancel each other out, and what is left is ____ positives.

3. Write an equation for the problem. Then solve it using guess-and-check or logical reasoning.

   Enrique purchased a set of acrylic paints for \$35 and fifteen brushes or p dollars each. He paid a total of \$80. What was the cost of one brush?

4. Write an inequality. Use negative integers where appropriate.

a. The show lasted at least 120 minutes.	**b.** Salt water remains frozen in temperatures below minus 2 centigrade.
c. Ashley owes on her credit card at least \$50.	**d.** Harry said, "The most you will owe me is \$12, no more."

5. Write using symbols, and simplify if possible.

   a. the opposite of the absolute value of 9

   b. the opposite of the difference $6 - 2$

   c. the absolute value of the opposite of 5

   d. the opposite of the sum $4 + 8$

Solve the equation $n \cdot n = n$. **Puzzle Corner**

Skills Review 9

1. Change each addition into a subtraction or vice versa. Then solve whichever is easier. Sometimes changing the problem will not make solving it easier, but the aim of this exercise is to practice making the change!

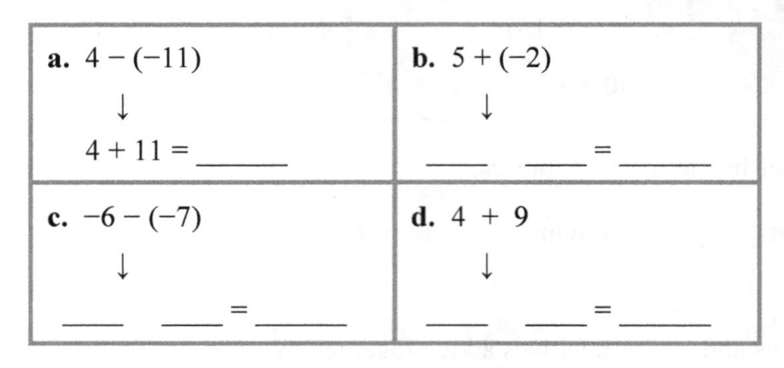

a. $4 - (-11)$ ↓ $4 + 11 =$ ______	**b.** $5 + (-2)$ ↓ ____ ____ = ______
c. $-6 - (-7)$ ↓ ____ ____ = ______	**d.** $4 + 9$ ↓ ____ ____ = ______

> ***Any* subtraction can be changed into an addition, *and* vice versa.**
>
> - Instead of subtracting a number, you can add its opposite.
> - Instead of adding a number, you can subtract its opposite.

2. Consider the expression $5 - x$. Is there a way to make the value of this expression to be *greater* than 5? If so, what kind of values of x will work for that?

3. Write a number sentence (actually, an equation!) to match the number line jumps.

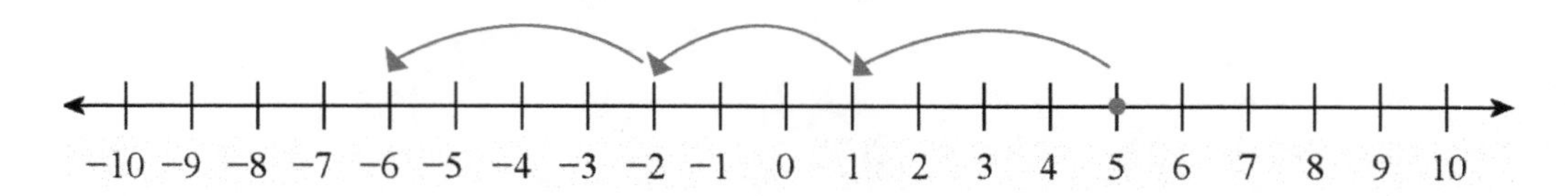

4. **a.** Draw the next steps for the pattern below.

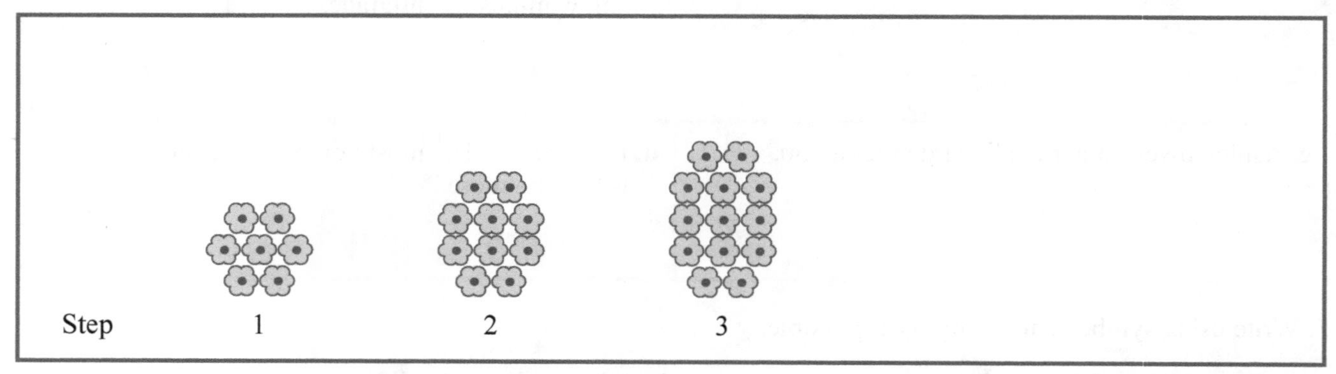

b. How do you see (visualize) this pattern growing?

c. How many flowers will there be in step 39?

d. What about in step n?

Skills Review 10

1. Divide in parts using mental math. You may end up with a fraction in the answer.

a. $\dfrac{800+1}{4}$	**b.** $\dfrac{30+1000}{5}$	**c.** $\dfrac{631}{6}$
d. $\dfrac{9x-3}{9}$	**e.** $\dfrac{a+14}{7}$	**f.** $\dfrac{4x+8}{8}$

2. Write a single mathematical expression ("number sentence") for each situation. Don't write just the answer.

a. There are m marbles in a container, and 1/3 of them are blue. Write an expression for the number of blue marbles. *blue marbles =*	**b.** There are m marbles in a container, and 1/3 of them are blue. Write an expression for the number of marbles that are not blue. *not blue marbles =*
c. The price of a printer is \$60, and it goes up by 1/10. Write an expression for the current price. *price =*	**d.** The price of a tablet costing p is discounted by 1/5. Write an expression for the discounted price. *price =*

3. Add and solve the riddle. *I'm tall when I'm young, and short when I'm old. What am I?*

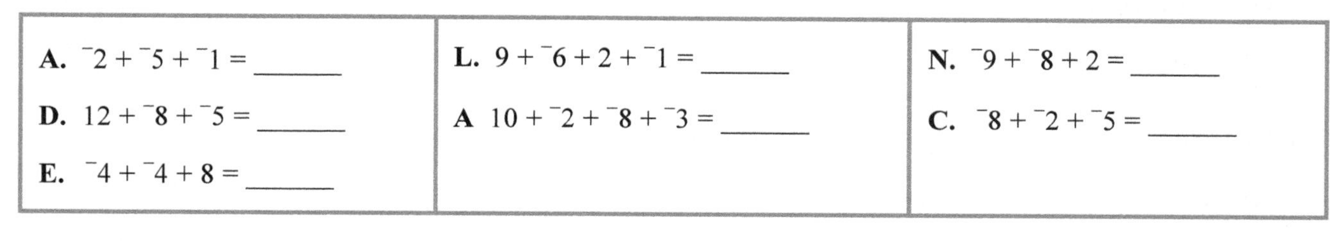

A. $^{-}2 + {}^{-}5 + {}^{-}1 =$ ______	**L.** $9 + {}^{-}6 + 2 + {}^{-}1 =$ ______	**N.** $^{-}9 + {}^{-}8 + 2 =$ ______
D. $12 + {}^{-}8 + {}^{-}5 =$ ______	**A** $10 + {}^{-}2 + {}^{-}8 + {}^{-}3 =$ ______	**C.** $^{-}8 + {}^{-}2 + {}^{-}5 =$ ______
E. $^{-}4 + {}^{-}4 + 8 =$ ______		

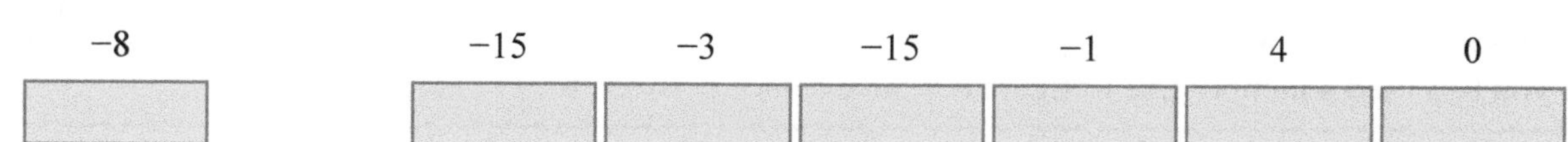

Some of the equations below will be true for *any* value of the variable n. They are said to have an infinite number of solutions. Find which ones they are.

$1 + n = 2.29$	$4 + n = 2 + 2 + n$	$5n + 40 = n + 30 + 4n$
$3n - 1 = 5$	$2n - 1 = n - 2 + n + 1$	$n = 0$

Skills Review 11

1. **a.** Find the value of the expressions $6 + a$ and $6 - a$ for the different values of a.

a	$6 + a$	$6 - a$
−5		
−4		
−3		
−2		

a	$6 + a$	$6 - a$
−1		
0		
1		
2		

a	$6 + a$	$6 - a$
3		
4		
5		
6		

b. For which values of a in the table is $6 - a$ more than $6 + a$?

c. For which value of a does the expression $6 - a$ have the value 8?

2. Let's say that m takes values from the set on the right.

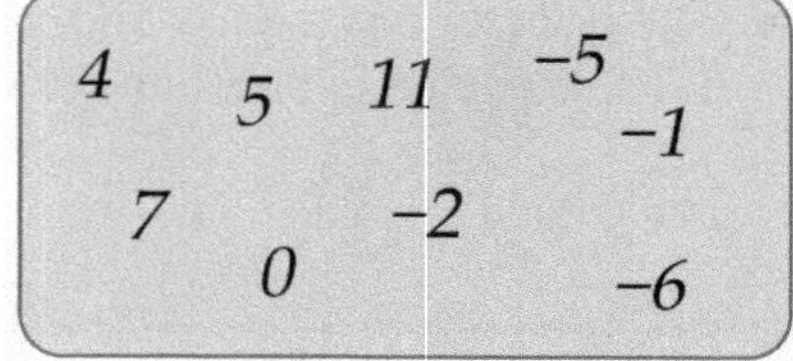

a. For which values of m would the expression $m + 8$ have the largest value?

The smallest?

b. For which values of m would the expression $8 - m$ have the largest value?

The smallest?

3. Which expressions can be used to find the distance between 24 and x?

a. $\vert -24 - x \vert$	**b.** $x - 24$	**c.** $\vert 24 - x \vert$	**d.** $24 - x$	**e.** $\vert x - 24 \vert$

4. Evaluate the expression $\vert a - b \vert$ for the given values of a and b. Check that the answer you get is the same as if you had used a number line to figure out the distance between the two numbers.

a. a is −2 and b is 9	**b.** a is −1 and b is −8
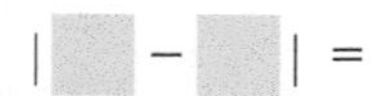 $\vert \square - \square \vert =$	

5. Circle the equation that matches the situation.

A sports club membership costs $30 per month, but now they are running a promotion for 25% (or 1/4) off of their normal prices.

If you have $400 to spare, how many months (m) of membership can you purchase?

$0.75(m - 30) = 400$	$30m - 0.25 = 400$
$0.75 \cdot 30m = 400$	$m(30/4) = 400$

Skills Review 12

1. Find the missing factors.

a. $-8 \cdot$ _______ $= -56$	**b.** $12 \cdot$ _______ $= -96$	**c.** $4 \cdot$ _______ $= -32$
d. $8 \cdot$ _______ $= 56$	**e.** $-12 \cdot$ _______ $= -96$	**f.** $-4 \cdot$ _______ $= -32$

2. Write an integer addition or subtraction equation to match these situations.

a. Missy is $50 in debt. Then she purchases a lunch for $12 on credit.

b. A group of tourists is at the Dead Sea, which is 430 m below sea level. Then they drive up the hill until they reach 350 m below sea level.

c. The temperature was 5°C but then it dropped seven degrees.

3. Subtract.

a. $2 - 16 =$	**b.** $-5 - (-8) =$	**c.** $40 - (-17) =$

4. Write "yes" or "no" to indicate if the operation is commutative or associative. Include one example using specific numbers.

Operation	Commutative?	Associative?	An example
addition			
subtraction			
multiplication			
division			

Shortcuts for simplification:	$- -$ can be changed to a single plus sign:	$n - (-m) = n + m$
	$+ -$ can be changed to a single minus sign:	$n + (-m) = n - m$
	$- +$ can be changed to a single minus sign:	$n - (+m) = n - m$

5. Apply the shortcuts from the box above to simplify these expressions by removing the parentheses.

a. $8x - (-2x) =$	**b.** $2y - (+7) =$	**c.** $7m + (-6n) =$
d. $2 + (-t) =$	**e.** $9s - (-s) =$	**f.** $u + (-u) =$

Skills Review 13

1. Match the equations with the situations and complete the missing parts.

   a. Liora was \$20 in debt. Then she bought a \$10 shirt on credit. Then she earned \$15. Now she has _____________.

   b. In a game, Matt had 20 negative points. Then he won a round and got 10 points. After that he gained another 15 points. Now he has _________ points.

   c. The temperature was −20°C. Then it rose 10°. Then it fell 15°. Now the temperature is ______ °C.

$-20 + 10 + 15 =$ ______
$-20 + 10 - 15 =$ ______
$-20 - 10 + 15 =$ ______

2. Add and subtract. It helps to change each subtraction into an addition.

a. $20 - (-3) - (-6) =$	**b.** $-9 - (-2) + (-3) =$
c. $-7 - 2 - (-3) + 4 =$	**d.** $5 + (-3) - (-9) - 2 =$

3. Evaluate the expressions. (Give your answer as a fraction or mixed number, not as a decimal.)

a. $\dfrac{a^2}{a-1}$, when $a = 5$	**b.** $\dfrac{2b-1}{b^2} - 1$, when $b = 3$
c. $\dfrac{5x^2}{x+1}$, when $x = 2$	**d.** $2 + \dfrac{9-y}{12-y}$, when $y = 0$

4. Here's a riddle for you. Solve the problems to uncover the answer.

O $2 \cdot (-9) =$ ______ **U** $-12 \cdot (-7) =$ ______ **A** $72 \div (-9) =$ ______

E $-24 \div (-12) =$ ______ **Y** ____ $\div\ 6 = -5$ **M** $-4 \cdot (-6) =$ ______

R ____ $\div\ (-6) = 2$ **N** $4 \cdot$ ____ $= -20$

What belongs to you but other people use it more than you?

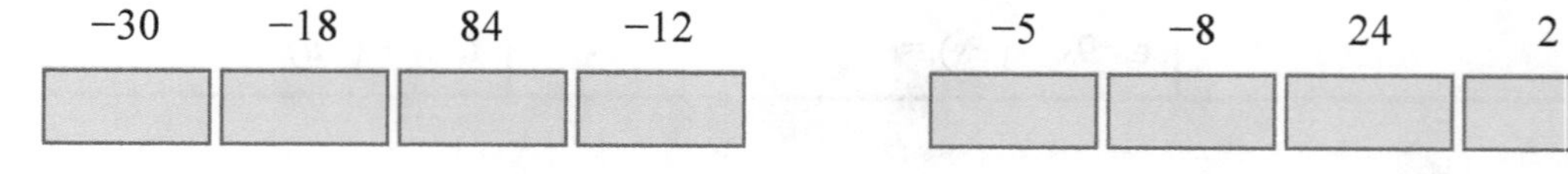

Skills Review 14

1. Which expressions can be used to find the distance between a and -7?

a. $a - 7$ **b.** $|a - (-7)|$ **c.** $|a - 7|$ **d.** $|-7 - a|$ **e.** $|-a - 7|$

2. Calculate the distance between s and -7 if...

a. s has the value -30 **b.** s has the value 30

3. Write as a fraction and simplify.

a. $6 \div (-4) = -\frac{6}{4} = -1\frac{1}{2}$	**b.** $14 \div (-21)$	**c.** $-32 \div 20$
d. $-2 \div (-8)$	**e.** $-48 \div (-64)$	**f.** $12 \div (-96)$

4. Write with symbols. Use a variable for the number.

a. The opposite of a certain number is more than negative 6.

b. The absolute value of a certain number is less than 5.

5. Write an expression with symbols. Then simplify.

a. The opposite of the sum five plus six

b. The absolute value of the product three times negative four

6. Each package of printer paper contains 500 sheets of paper, and the package costs \$4.50. Write an expression for the cost of s sheets of paper.

7. Simplify the expressions.

a. $2x - 8x - 9$	**b.** $n^3 + 8n^3 - 3n^3$	**c.** $2s^2 - 8s^2 - 8s^2 - 3$

Printing at a print shop costs \$0.08 per page. Sheila has a coupon that will give her a 10% discount off of her purchase.

a. Write an expression for her total cost, if she needs to print p pages.

b. How many pages can she print with \$14, using her coupon?

Skills Review 15

1. Find the value of the expressions using the correct order of operations.

a. $4 + 8 - 5 \cdot (-8)$	**b.** $(3 + (-4)) \cdot 9$	**c.** $4 - 5 \cdot 7 + 6$
d. $-1 + \dfrac{1}{2 - 5}$	**e.** $8 - \dfrac{1}{-4}$	**f.** $\dfrac{-12}{4} \cdot 2 + 9$

2. Factor these expressions (write them as products) using the distributive property "backwards".

a. $54x + 36 =$	**b.** $32x - 40y + 8 =$
c. $45a - 85b + 50 =$	**d.** $28 + 49a - 77b =$

3. A computer game adds 300 bonus points to your initial point count when you complete a level. Also, if you finished in less than 4 minutes, your points are doubled.

   a. Write an expression for your final point count if you finish a level in less than 4 minutes. Use p for the initial point count (before the bonus points and doubling).

   b. Let's say someone finishes a level in 3 minutes with lots of errors, and ends up with −200 points initially. What will be their final point count?

   c. Let's say Eric finishes a level quickly and the computer calculates that he gets −100 points. What was his initial point count? Write an equation for this situation. Then solve it using guess and check or logical thinking.

4. The area of a two-part rectangle is given by the equivalent expressions $m(x + 8)$ and $mx + 8m$. Sketch the rectangle and mark the side lengths m, x, and 8 in it.

5. Find the missing numbers. Use guess and check and logical thinking.

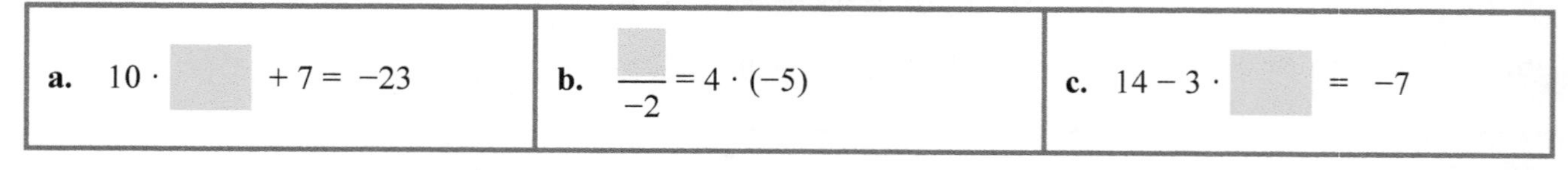

a. $10 \cdot \square + 7 = -23$	**b.** $\dfrac{\square}{-2} = 4 \cdot (-5)$	**c.** $14 - 3 \cdot \square = -7$

Skills Review 16

1. How much more money does Roy have than Rob, if Rob's balance is −\$71 and Roy's is −\$15? Write an expression using absolute value, and calculate its value.

2. Is it possible to calculate the distance between two numbers m and n as $n - m$ instead of $|n - m|$? Justify your answer by giving examples that use specific numbers.

3. The table gives the average high and low temperatures for all 12 months for Portland, Maine.

   a. Calculate the difference between the two, for each month.

   b. When is the difference between the high and low temperatures the greatest?

Months	High (C°)	Low (C°)	DIFFERENCE
January	−1	−11	
February	2	-9	
March	6	−4	
April	12	2	
May	18	7	
June	23	12	
July	26	15	
August	26	14	
September	21	10	
October	15	4	
November	9	−1	
December	3	−7	

4. Fill in the missing numbers.

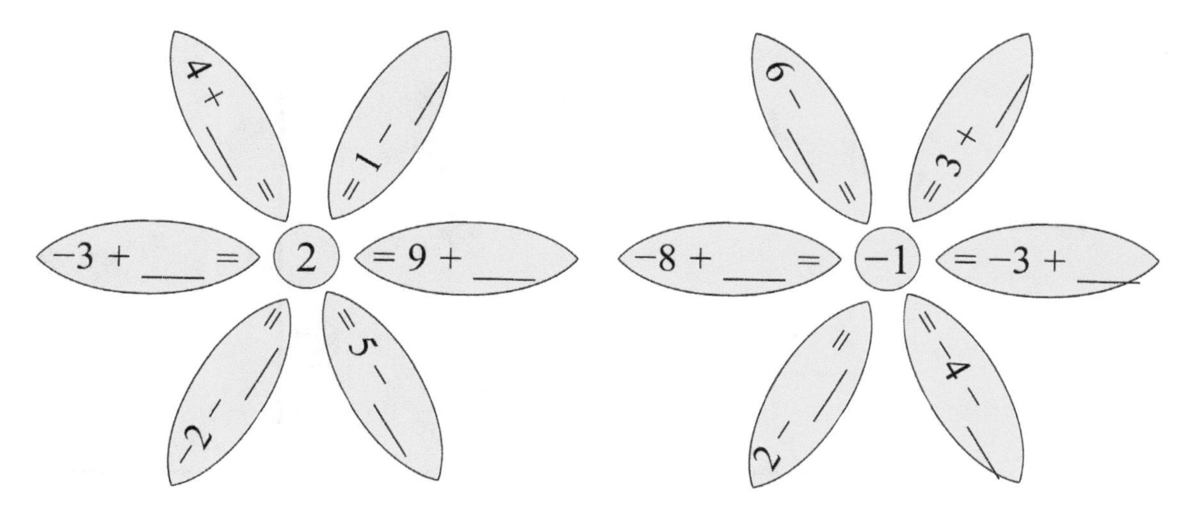

Skills Review 17

1. Write an equation, then solve it using guess and check. Each root is between −20 and 20.

a. 8 less than a equals −1.	**b.** Six equals the quantity b plus 1 divided by 3.
c. Four times the quantity y minus 1 equals 12.	

2. Multiply.

a. $-2 \cdot 8 \cdot (-2)$

b. $(-4) \cdot (-1) \cdot (-3) \cdot (-5)$

c. $3 \cdot (-5) \cdot (-10) \cdot (-2) \cdot 10$

d. $7 \cdot (-2) \cdot (-4) \cdot (-1)$

3. What is the total value, in cents, if you have x nickels, y dimes and s quarters? Write an expression.

4. Solve the equation. Write in the margin what operation you do to both sides.

Balance	Equation	Operation to do to both sides
↓ ↓	$2x + 3 = -1$	

Skills Review 18

1. Solve. Check your solutions.

a. $-10 = x + 6$	**b.** $-11 = w - 20$
c. $-t = 2 + (-5)$	**d.** $8 - y = -6$

2. Simplify the expressions.

a. $m - 7 + m + m$	**b.** $m \cdot 7 \cdot m \cdot m \cdot m$	**c.** $5y \cdot 2y \cdot 3 \cdot y \cdot y \cdot 3$

3. **a.** Write an expression for the area of a rectangle with sides $6r$ and $3r$.

b. What is the perimeter of that rectangle?

4. Draw counters to illustrate the sum $4 + (-2) + (-5) + 2$.

5. Write an integer division where the quotient is less than 0, and both the dividend and divisor are divisible by 4.

6. Add and subtract. Check: the sum of the four *answers* is zero.

a. $4 - (-3) - (-5) + 1 =$	**b.** $-5 + (-9) - (-5) - 2 =$
c. $-9 - 6 - (-6) + 4 =$	**d.** $2 + (-2) - (-7) - 4 =$

Skills Review 19

1. Solve. Check your solutions.

a. $\dfrac{x}{8} = -8 + (-7)$	**b.** $-\dfrac{1}{5}x = -20 + 10$

2. Solve. Check your solutions.

a. $\dfrac{1}{9}x = -9$	**b.** $-100 = \dfrac{s}{-8}$	**c.** $-3 = -\dfrac{1}{11}x$

3. Simplify.

a. $-|40|$ **b.** $|-5| - |2|$ **c.** $-|-11|$ **d.** $20 - |-5|$

Step 1 2 3 4 5 6

4. **a.** Draw the next steps.

b. How do you see this pattern growing?

c. How many flowers will there be in step 39?

d. What about in step n?

Skills Review 20

1. On a particular winter day in Oslo, Norway, the highest temperature was 3°C and the lowest was −8°C.

   a. What was the difference between the highest and the lowest temperature?

   b. Write an *expression* for the difference you just calculated.

2. For each given situation, write an equation and solve it. The problems themselves are simple, and you could solve them without writing an equation, but it is important to practice writing equations for these simple situations now, so you will be able to write equations for more complex situations later on.

   a. If each web page lists 20 search results, then how many pages do you need to list a total of 1800 search results?

   b. Seth had an 8-foot piece of wood. He cut it into two pieces. One piece measured 3 1/4 ft. How long was the other piece?

3. **a.** Mark $-1\frac{3}{8}$ on the number line.

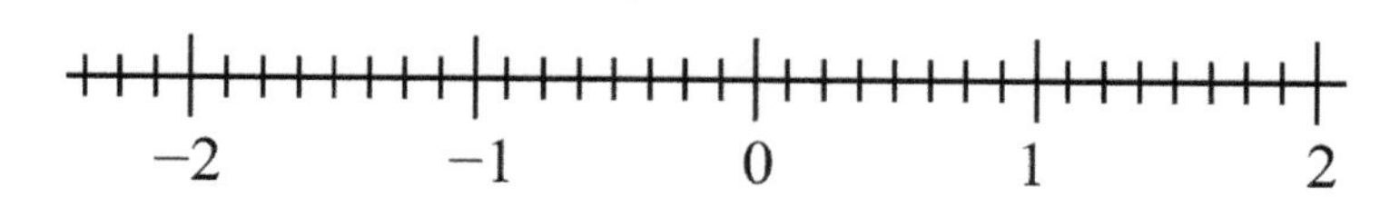

   b. Make a *division* problem that has an answer of $-1\frac{3}{8}$.

4. Use the formula $d = vt$ to solve the problems.

a. A Canadian goose flies a distance of 50 miles in 1 hour 15 minutes. What is his average speed? d = v t ↓ ↓ ↓	**b.** If a boy can bicycle at a constant speed of 9 km/h, how long will it take him to cover a distance of 15 km? d = v t ↓ ↓ ↓

Skills Review 21

1. **a.** Which expressions below give you the volume of a cube with edges 3 cm long?

(i) 3 cm^3 **(ii)** $(3\text{ cm})^3$ **(iii)** 27 cm^3 **(iii)** 9 cm^3

b. Write an expression for the volume of a cube with edges that are *s* units long.

c. What is the edge length of a cube with a volume of 1000 m^3?

2. Explain (such as to your friend) how to add one negative and one positive number, when they are both 3-digit numbers, such as 293 and −714.

3. Write the fractions and mixed numbers marked by the arrows.

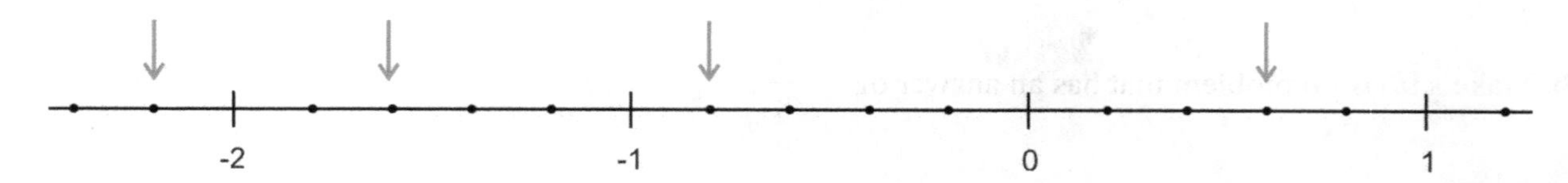

4. Name one operation that is commutative and one that is not. Give numerical examples for both.

5. Write these numbers as a ratio (fraction) of two integers.

a. 3	**b.** −87	**c.** 0.74	**d.** −2.1

6. Form a fraction from the two given integers. Then convert it into a decimal.

a. −8 and 6	**b.** 10 and −25	**c.** −42 and −100

Skills Review 22

1. Write a multiplication expression for the repeated pattern in the model.
   Then multiply the expression using the distributive property.

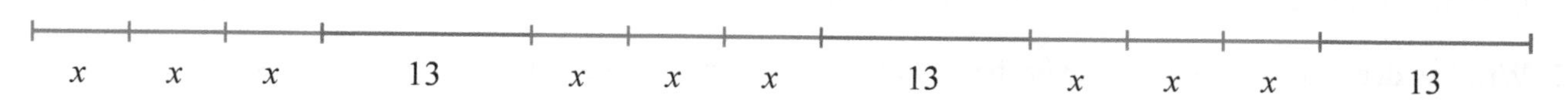

2. Factor these expressions (write them as products). Think of divisibility!

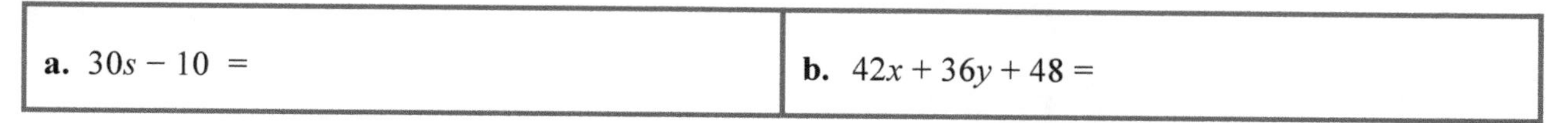

a. $30s - 10 =$	**b.** $42x + 36y + 48 =$

3. A regular chessboard has 8 × 8 squares. Consider an $n \times n$ chessboard from which two corner squares are removed.

   a. Write an expression for the area of an $n \times n$ chessboard from which two squares are removed.

   b. For which value of n would the expression you wrote for part (a) have the value of 119?

4. **a.** Make up a shopping situation to match this expression: $10(\$x - \$1.25)$

   b. Multiply the expression using the distributive property.

5. Solve. Check your solutions.

a. $8 = -12 - x$	**b.** $11 - x = 7 - 13$
c. $-5 + y = -3 + 21$	**d.** $2 + (-9) = 1 - z$

Skills Review 23

1. **a.** Is 0.352352352... = $0.\overline{352}$ a rational number?

   b. Which is more, $0.\overline{5}$ or 0.5? How much more?

2. Write as decimals, using a line over the repeating part (if any). Use long division if necessary.

a. $3\frac{7}{12}$	**b.** $\frac{11}{30}$	**c.** $-\frac{1}{20}$

3. Add the fractions.

a. $\frac{1}{5} + \left(-\frac{7}{8}\right)$	**b.** $-\frac{1}{6} + \frac{3}{7}$

4. Draw a number line jump for each subtraction.

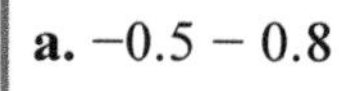

a. $-0.5 - 0.8$ **b.** $1 - 2.4$ **c.** $-2.1 - (-0.5)$

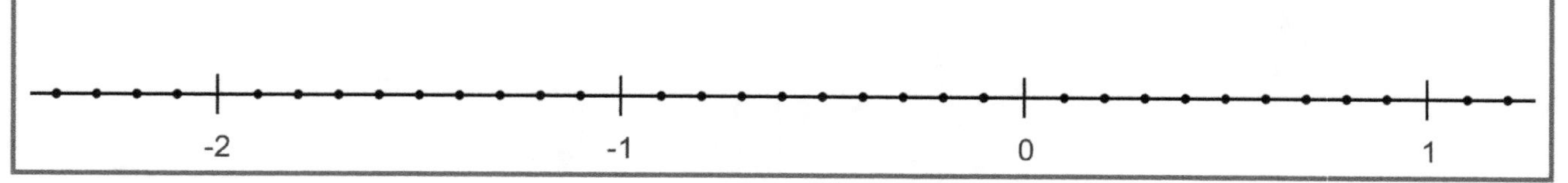

Skills Review 24

1. Write the rational numbers in their four forms. Write the fraction in lowest terms.

	ratio		fraction		decimal		percent
a.		=	$\frac{11}{25}$	=		=	
b.		=		=		=	6%

2. Multiply these in your head.

a. $(-0.1)^2$	**b.** $(-0.2)^3$	**c.** $-0.11 \cdot 0.02$
d. $0.7 \cdot (-1.2)$	**e.** $-0.8 \cdot 0.1 \cdot (-0.3)$	**f.** $-0.5 \cdot (-0.005)$

3. Multiply.

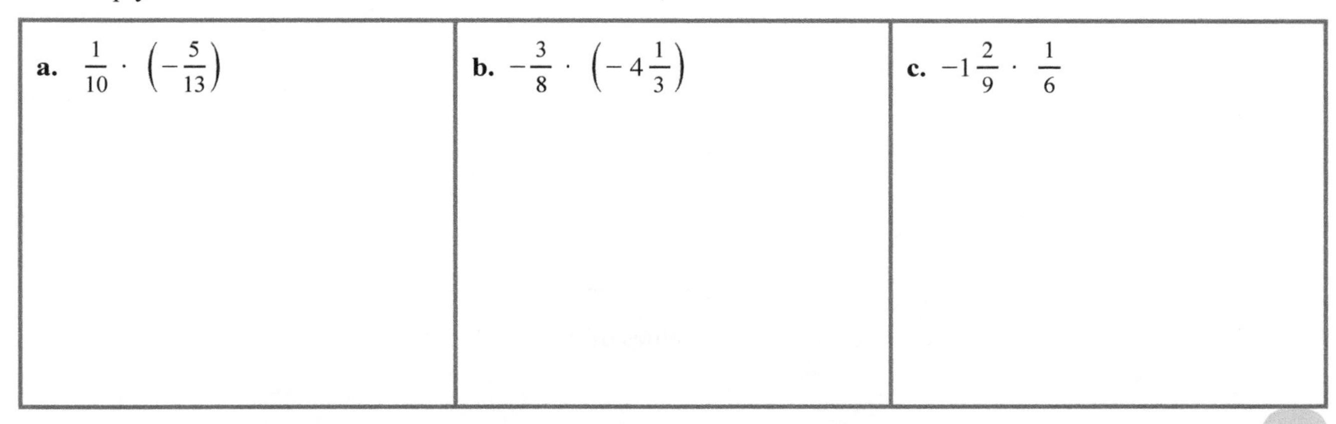

a. $\frac{1}{10} \cdot \left(-\frac{5}{13}\right)$	**b.** $-\frac{3}{8} \cdot \left(-4\frac{1}{3}\right)$	**c.** $-1\frac{2}{9} \cdot \frac{1}{6}$

4. Add and subtract. The answers are on the right.

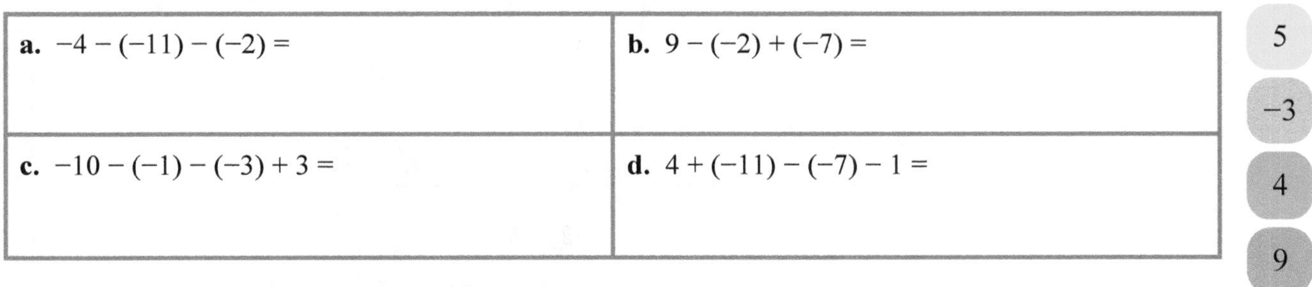

a. $-4 - (-11) - (-2) =$	**b.** $9 - (-2) + (-7) =$
c. $-10 - (-1) - (-3) + 3 =$	**d.** $4 + (-11) - (-7) - 1 =$

−1 5 −3 4 9

5. Simplify the expressions.

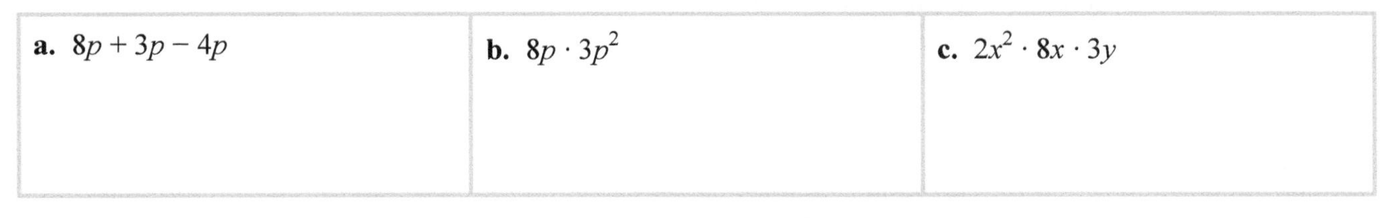

a. $8p + 3p - 4p$	**b.** $8p \cdot 3p^2$	**c.** $2x^2 \cdot 8x \cdot 3y$

Skills Review 25

1. Multiply.

a. $14.2 \cdot (-0.094)$	b. $-2.8 \cdot (-0.0005)$	c. $-0.8 \cdot 0.002 \cdot (-0.4)$

2. Divide.

a. $-\frac{7}{8} \div \left(-\frac{1}{2}\right)$	b. $-12 \div \frac{5}{4}$

3. **a.** Find the value of the expression $3/x$ for different values of the variable x.

x	$\frac{3}{x}$
6	
5	
4	

x	$\frac{3}{x}$
3	
2	
1	

x	$\frac{3}{x}$
0	
−1	
−2	

x	$\frac{3}{x}$
−3	
−4	
−5	

b. Find a value of x so that $3/x$ has the value $-1/2$.

Skills Review 26

1. Write the numbers with scientific notation.

a. 506,000 **b.** 34,099,000

c. 17,200 **d.** 644,500,000

2. Write with symbols. Use a variable for the number. Then give two numerical values that fit the condition.

a. The absolute value of a certain number is less than 6.

Example values:

b. The opposite of a certain number is less than 2.

Example values:

c. The opposite of a number is greater than minus 5.

Example values:

3. Solve. Check your solutions.

a. $\frac{1}{8}x = -20$	**b.** $-100 = \frac{1}{9}x$	**c.** $-\frac{1}{2}x = 4 + 8$

4. Write an equation and solve it. Do not simply solve the problem without writing an equation. First think: what thing is unknown? Use a variable to denote that.

Joan is riding a distance of 65 km for exercise. She has bicycled at a steady speed of 10 km/h for some time, and now has 30 km left to go. How long has she been bicycling at this point?

Equation:

Skills Review 27

1. Solve.

a. $2\frac{3}{8} - x = \frac{1}{2}$	**b.** $1\frac{1}{5} + v = \frac{3}{10}$

2. Choose the correct way to enter these calculations into a calculator and solve the expressions.

a. Calculation: $\frac{4}{11} \cdot 24.06$	**b.** Calculation: $120 \div \frac{7}{8}$	**c.** Calculation: $156 \cdot \frac{29}{100}$
1. $4 \div 11 \cdot 24.06$	1. $120 \div 7 \div 8$	1. $156 \cdot 29 \div 100$
2. $4 \cdot 24.06 \div 11$	2. $120 \div (7 \div 8)$	2. $156 \cdot (29 \div 100)$
3. It does not matter; both will give the correct answer.	3. It does not matter; both will give the correct answer.	3. It does not matter; both will give the correct answer.

3. Make many different division and multiplication problems with an answer of negative four. Try to be creative!

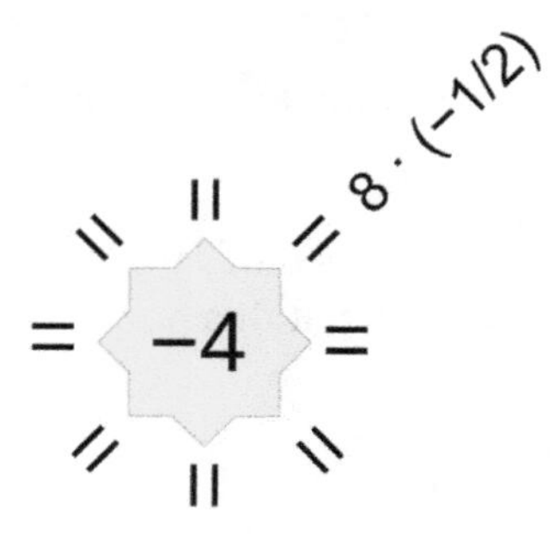

4. If a boat travels at a constant speed of 12 km/h, how long will it take to travel a distance of 5.6 km?

Skills Review 28

1. Simplify.

a. $2x - (-t) =$	**b.** $-9 - (-2x) =$	**c.** $-2w + (-3w) =$
d. $2s + (-3s) =$	**e.** $8 - (+7x) =$	**f.** $7h + (-10g) =$

2. Solve.

a. $8x = -\frac{3}{4}$	**b.** $\frac{z}{8} = -\frac{11}{12}$

3. Solve using *fraction* arithmetic.

a. $\frac{1}{3} - 1.2 - \frac{3}{4}$	**b.** $0.4 \cdot \frac{5}{6} - 2\frac{1}{2}$

Skills Review 29

1. **a.** Draw the next two steps.

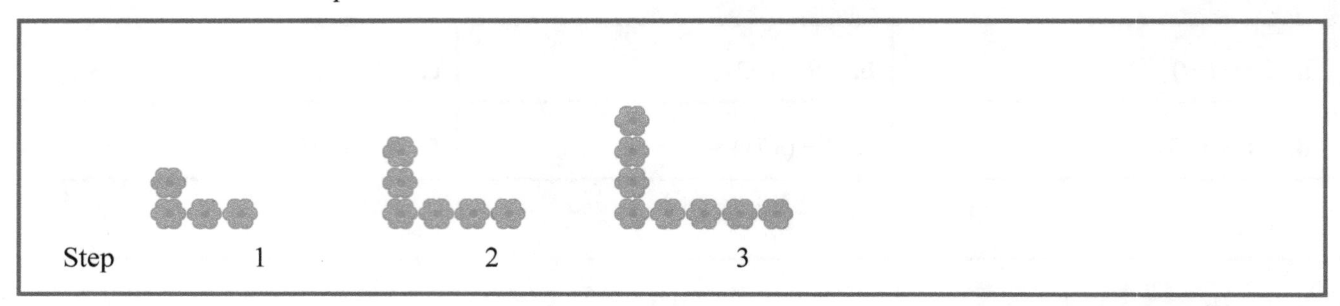

b. How do you see this pattern growing?

c. How many flowers will there be in step 39?

d. What about in step n?

2. Solve the equations. Check your solutions.

a. $0.76 = x - 0.14$	**b.** $2 - y = 0.26$
c. $\frac{x}{8} = -2.15$	**d.** $-0.99x = -1.05$

3. Write these decimals as mixed numbers.

a. 5.0009	**b.** −37.3920483	**c.** 60.00605

4. Write the fractions as decimals.

a. $-\frac{589}{10{,}000}$	**b.** $\frac{2{,}055}{10}$	**c.** $\frac{40{,}954}{1{,}000{,}000}$

Skills Review 30

1. Write an equation for each situation. Think what thing is unknown, and choose a variable for that. You do not have to solve the equations.

   a. The perimeter of a square is 5.6 m. How long is its side?

   b. Will is 15 years younger than Kyle. If Will is 56, how old is Kyle?

2. Solve. Check your solutions.

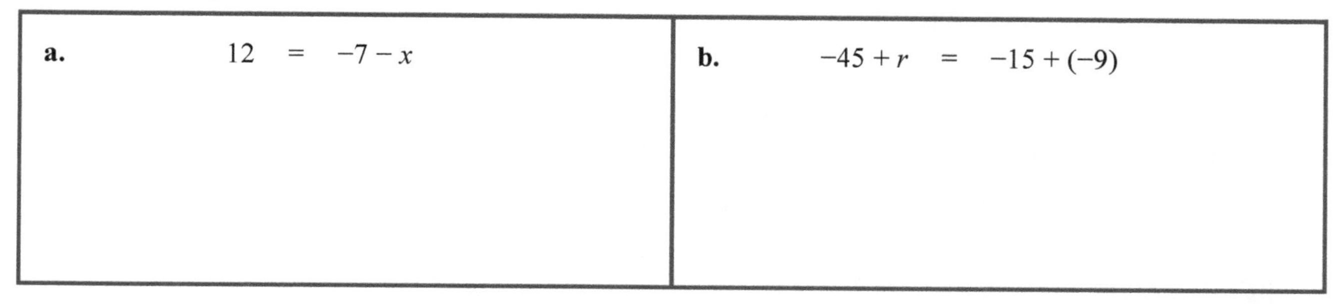

a. $12 = -7 - x$	**b.** $-45 + r = -15 + (-9)$

3. Which expression(s) can be used to find the distance between −13 and −41?

   a. $-13 - 41$ **b.** $|-13 + (-41)|$ **c.** $|-41 - (-13)|$ **d.** $13 - 41$ **e.** $|-41 + 13|$

4. Which expression(s) can be used to find the distance between a and 7?

   a. $|7 - a|$ **b.** $a - 7$ **c.** $|a - 7|$ **d.** $|a + 7|$ **e.** $|a - (-7)|$

5. Place operation symbols in the empty spaces so that the equation will be true.

a. $6 + 3 \;\square\; (-5) = -9$	**b.** $-24 \;\square\; 6 + 4 \;\square\; 1 = 0$	**c.** $(-5 \;\square\; 4) \;\square\; 8 \;\square\; 2 = -10$

6. Solve.

a. $-\frac{1}{4} - \left(-\frac{1}{2}\right) + \frac{5}{8}$	**b.** $\frac{2}{3} + \left(-\frac{5}{6}\right) + \frac{1}{4} + \left(-\frac{1}{2}\right)$

Skills Review 31

1. Solve. Check your solutions (as always!).

a. $\dfrac{x-6}{7} = 5$	**b.** $\dfrac{y+20}{-7} = 1$
c. $9 - 3h = 3$	**d.** $22 = 10 - 4t$

2. Divide in parts using mental math. You may end up with a fraction in the answer.

a. $\dfrac{8{,}000 + 400 + 7}{8}$	**b.** $\dfrac{4{,}803}{4}$	**c.** $\dfrac{691}{3}$
d. $\dfrac{2x-1}{2}$	**e.** $\dfrac{49a+7}{7}$	**f.** $\dfrac{-4b+2}{4}$

3. A barium ion has 56 protons and 54 electrons. The electric charge of the protons is $56e$, and the charge of the electrons is $-54e$. What is the total electric charge of this ion?

4. Solve.

a. $6.4 + (-0.7) - (-1.9)$	**b.** $2.08 - 3.6$	**c.** $-1.01 - 0.908$

Skills Review 32

1. **a.** Find the value of the expression $-6 - 2x$ for at least six different values of x. Choose values near zero, and organize your work in a table.

   b. For which value of x will the expression $-6 - 2x$ have a value of 0?

2. Circle the equations that match the situation.

   a. The price of a book is discounted by 1/5, and now it costs $12.

$p - \frac{p}{5} = \$12$	$\frac{p}{5} = \$12$	$\frac{5p}{4} = \$12$	$\frac{4p}{5} = \$12$	$p - 1/5 = \$12$

   b. Ann took a $3000 loan and now she is paying it back with monthly payments of $75. After how many payments will she have $600 left to pay?

$\frac{3000 - 600}{75} = x$	$\frac{3000 - x}{75} = 600$
$75x + 600 = 3000$	$75(x + 600) = 3000$

3. Give a real-life context for each multiplication. Then solve.

a. 1.08 · $315
b. (9/10) · 160 lb
c. (1/2) · (3/4)

Skills Review 33

1. Fill in the pyramid! Add each pair of terms in neighboring blocks and write its sum in the block above it.

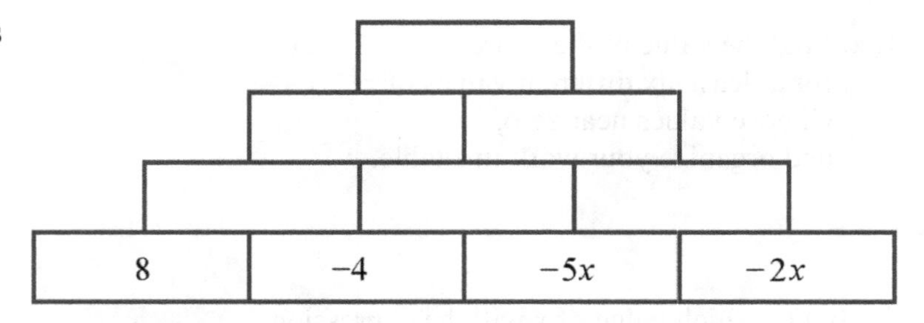

2. Solve. Check your solutions.

a. $-6x + 2x = 4x - x - 6$	b. $4 + 50y - 10y = 10 - 20y$

3. Draw a bar model for the equation $12.6 + 0.9 + w + 91.9 = 290$, and solve it.

4. Are the expressions equal, no matter what value n has? If so, you don't need to do anything else. If not, provide a counterexample: a specific value of n that shows the expressions do NOT have the same value.

a. $(n - 10) \cdot 7$ $7(n - 10)$	b. $\dfrac{5}{n+1}$ $\dfrac{n+1}{5}$

Skills Review 34

1. A section of a flower garden has rows of flowers. The first row has three flowers, and each row after that has two more flowers than the previous row.

1
2
3
...

Row	Flowers
1	3
2	3 + 2
3	
4	
n	

a. Write a formula that tells the gardener the number of flowers in row n.

b. How many flowers are in the 28th row?

c. In which row will there be 97 flowers?

2. Write an expression with *three* terms: one has variable part x and coefficient 2, another is a constant equal to one dozen, and yet another has variable part y and coefficient 5.

3. Find the missing factors.

a. $7 \cdot \_\_\_\_\_\_ = -56$	**b.** $-9 \cdot \_\_\_\_\_\_ \cdot 2 = 108$	**c.** $5 \cdot 3 \cdot \_\_\_\_\_\_ = -600$

4. Write an inequality. Use negative integers where appropriate.

a. Mary owes the bank at least $2,000.

b. You have to be at least 21 to buy this.

c. This medicine should be kept at 5° or colder.

5. Solve.

a. $-10x = \frac{9}{8}$	**b.** $\frac{4}{5} - x = \frac{2}{3}$

Skills Review 35

1. The chart shows the average minimum temperatures for Toronto, for six months. Calculate their average...

Oct	Nov	Dec	Jan	Feb	Mar
7	2	−4	−7	−6	−2

a. ...as a fraction or mixed number.

b. ...as a decimal, rounded to two decimal digits.

2. Use the distributive property to multiply.

a. $20(t - 6) =$	**b.** $0.5(a + 2b) =$	**c.** $7(3x + 2y + 0.4) =$

3. Solve mentally! (Hint: the distributive property will help.)

a. $6 \cdot 99 =$	**b.** $4 \cdot 98 =$	**c.** $7 \cdot 299 =$

4. Solve. Check your solutions.

a. $-\frac{1}{2}x = -24$	**b.** $-14 = \frac{1}{10}x$	**c.** $\frac{1}{4}x = 18 - 22$

d. $\frac{3}{5}z = 120$	**e.** $\frac{4}{9}y = 60$

Skills Review 36

1. Solve the inequalities and plot their solution sets on a number line. Write appropriate multiples of ten under the bolded tick marks (for example, 30, 40, and 50).

a. 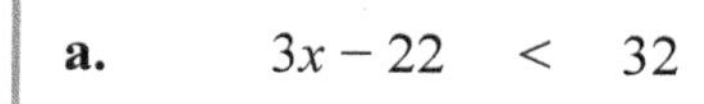 $3x - 22 \quad < \quad 32$ 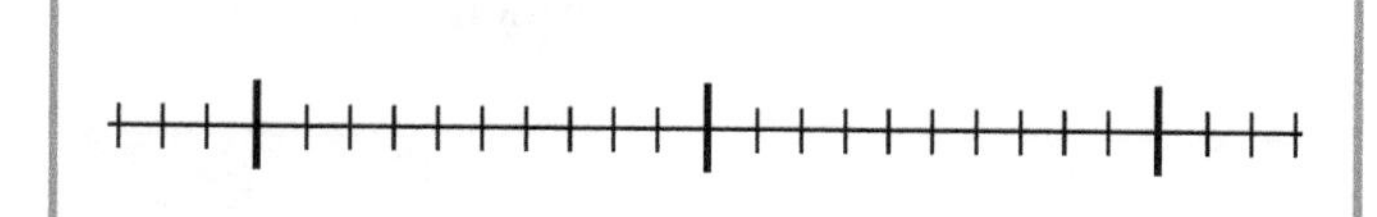	**b.**  $-50 \quad \leq \quad 10x + 30$

2. Write an equation to match the situation, and simplify it.

The temperature was −11°C and fell 3°. Then it rose 7°. Now the temperature is ______ °C.

Equation:

3. Add and subtract.

a. $-40 + (-13) - (-12) =$	**b.** $0 - (-2) - (-9) =$
c. $-9 + (-8) + 10 + (-8) =$	**d.** $14 - (-21) - 7 - 3 =$

4. How long will it take you to drive a distance of 70 miles, if you drive 1/5 of it in rush-hour traffic, at the speed of 35 mph, and the rest with a speed of 50 mph?

5. Simplify these complex fractions.

a. $\dfrac{2}{\frac{5}{9}}$	**b.** $\dfrac{\frac{3}{8}}{\frac{1}{12}}$	**c.** $\dfrac{\frac{4}{5}}{-10}$

Skills Review 37

1. Write an <u>inequality</u> for this problem. And solve the problem.

At a fairgrounds, Sheila is planning to ride the roller coaster twice ($5.50 per ride), and then SuperFall ($3.50 per ride) as many times as she can afford. If she has $28 to spend, how many times can she ride SuperFall?

2. Solve each problem using an equation and also using some other strategy, such as a bar model or mental reasoning.

Three-eighths of a number is 12.45. What is the number?	
<u>Equation:</u>	<u>Another way:</u>

3. Write the numbers with scientific notation.

a. 560,000

b. 3,290,000

c. 40,100

d. 24,500,000

4. Complete the "cross-number puzzle".

Across:

a. $2 \cdot (-13)$

b. $-4 \cdot (-8)$

d. $80 \cdot (-7)$

e. $-120 \div (-3)$

f. $4{,}000 \div$ _____ $= (-8)$

Down:

a. _____ $= -42 \div 2$

c. $-96 \div (-4)$

d. $-1{,}000 \div 20$

e. _____ $\div (-14) = 5$

a.			b.	c.
	d.			
e.				e.
		f.		

Skills Review 38

1. **a.** Plot the equation $y = 2x - 1$.

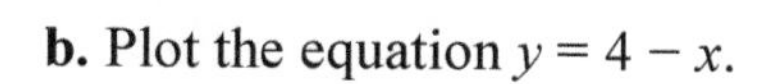

   b. Plot the equation $y = 4 - x$.

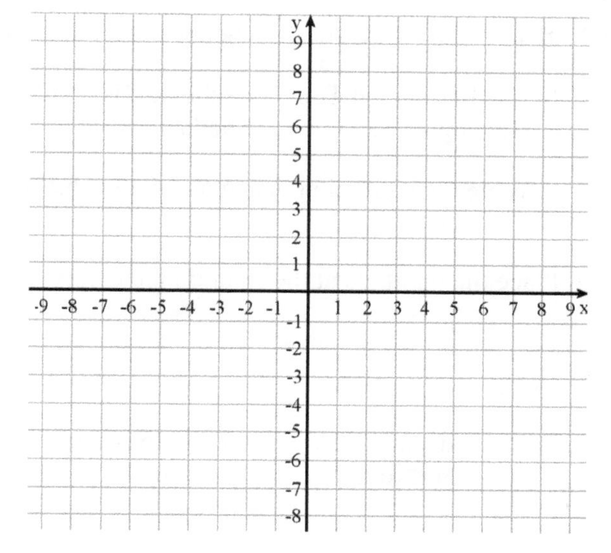

2. Determine the slopes of the above lines.

3. Explain two different ways to determine if the point $(2, -3)$ is on the line $y = 4 - x$.

4. Place parenthesis in this expression in TWO different ways, so that you get TWO different values for it.

(1) $-8 + 5 \cdot 3 - 12$	**(2)** $-8 + 5 \cdot 3 - 12$

5. Solve.

a. $31 = 5 - 2x$	**b.** $3 - 4t = -13$
c. $\frac{x}{5} + 3 = -5 \cdot (-3)$	**d.** $\frac{-3x}{4} = -21$

Skills Review 39

1. **a.** Write an expression, using absolute value, to find the distance between x and y.

   b. The balance on Tony's credit card is −\$200, and the balance on Sheila's is \$40. How much better off is Sheila?

   c. Write an expression, using absolute value, to match the question in (b).

2. Place the numbers at the right on the empty lines to make true equations.

   a. _____ + _____ + _____ = −8

   b. _____ + _____ + _____ = 0

   2   −11   6   −3   −7   5

3. Write in decimal form. Use long division. If you find a repeating pattern, give the repeating part. If you don't, round your answer to five decimals.

a. $1\frac{7}{13}$	**b.** 0.98 ÷ 6

4. Plot each line and find its slope.

   a. $y = (1/2)x$

   slope: _______

   b. $y = -2x - 1$

   slope: _______

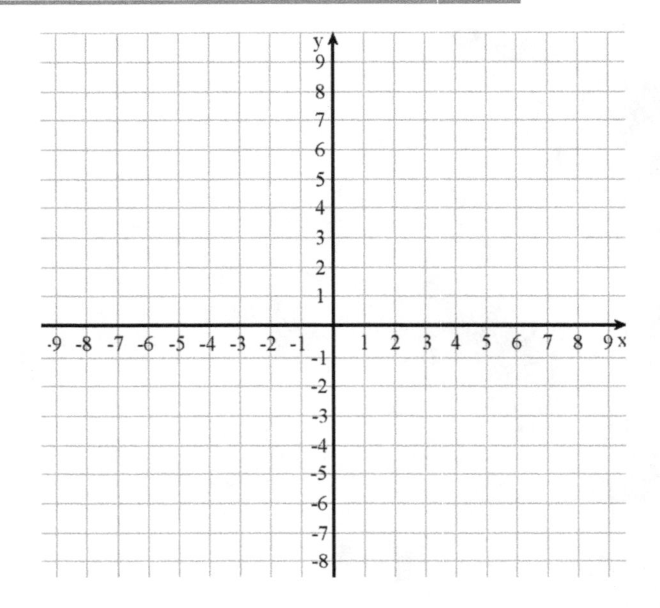

Skills Review 40

1. **a.** Find the value of the expression $3x - 2$ for at least six different values of x. Organize your work in a table. You should see a pattern if you choose the values of x that are in some order.

   b. For which value of x will the expression $3x - 2$ have a value of 0?

2. The tables below show the time (t) and distance (d) of two people running with a constant speed. Continue the pattern in each table. Write equations that relate t and d, and plot the equations.

a.

t	0	1	2	3	4	5	6	7
d	0	8	16	24				

equation: ______________________

b.

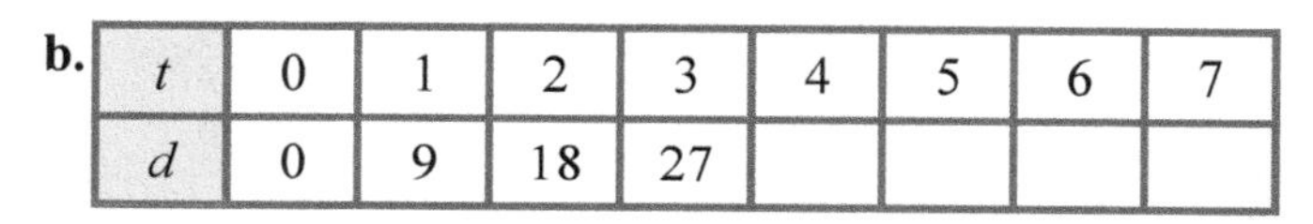

t	0	1	2	3	4	5	6	7
d	0	9	18	27				

equation: ______________________

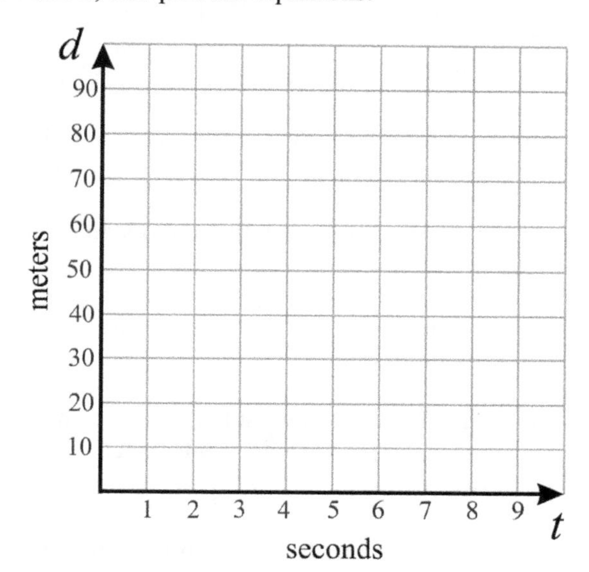

c. How much is the second runner ahead of the first when $t = 10$ s?

3. A bus company needs to increase their prices because costs have increased. Currently, on a certain route, people pay $2.25 for a single ride. On average, 420 people ride that route every day. By how much should they increase their price, in order to simply cover the costs ($1,050 per day)? Choose an equation that matches the problem, and solve it.

$420x + 2.25 = 1050$	$1050 - 420 \cdot 2.25 = x$
$420(2.25 + x) = 1050$	$420x + 2.25x = 1050$

Skills Review 41

1. Two bicyclists, Henry and Andy, start at the same time bicycling through a long route, and stay together for the first hour.

a. Plot the points from the tables below.

Henry

Distance (km)	0	10	20	30	40	50	60
Time (h)	0	0.5	1	1.5	2	2.5	3

Andy

Distance (km)	0	10	20	28	36	44	52
Time (h)	0	0.5	1	1.5	2	2.5	3

b. What is Henry's average speed during the first hour? Second hour?

c. Write an equation relating the distance Henry covers and the time it takes him.

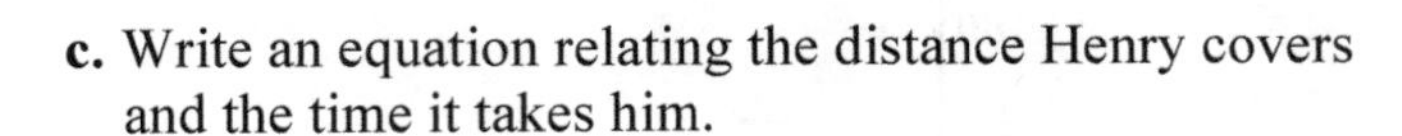

d. What is Andy's average speed during the first hour? Second hour?

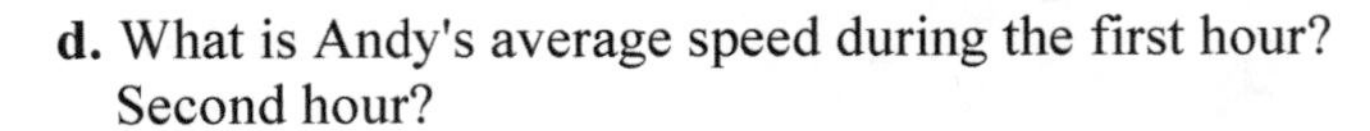

e. If they both continue at the constant speeds they have been going for one more hour, how much further will Henry be at 4 hours than Andy?

2. Use a fraction line to write ratios of the given quantities. Then simplify the ratios to integers.

a. \$1.60 and \$3.00	**b.** 1.4 km and 1.8 km
c. ¼ mi and 1 ½ mi	**d.** ⅔ L and 1 L

3. Solve.

a. $-1.34 - 5.9$	**b.** $0.6 + (-0.9)$	**c.** $-45.6 + 23.8$

4. Solve. Check your solutions.

a. $6x + 3x + 1 = 9x - 2x - 7$	**b.** $16y - 4y - 3 = -4y - y$

Skills Review 42

1. Solve. Check your solutions.

a. $-2x + 3x + 5 = 11x - 5x - 3$	**b.** $40 + 20b - 4b = 2b + 50 - 7b$

2. Divide. In (d), give your answer to three decimal digits.

a. $-4 \div \frac{8}{9}$	**b.** $-\frac{11}{12} \div \left(-\frac{1}{3}\right)$
c. $2.7 \div 0.004$	**d.** $54.8 \div 0.11$

Skills Review 43

1. Meg earns \$45 for three hours of work.

   a. In how many hours will she earn \$750?

   b. If she works three hours a day, how many (full) work-days will it take her to earn at least \$750?

2. Solve. Check your solutions (as always!).

a. $50x - 16 = -10 - 30x$	**b.** $-2k + 8 = 40 - 15k$

3. Simplify by removing the parentheses. Then solve.

a. $8 + (-2)$	**b.** $-5 - (-6)$	**c.** $-10 + (-40)$	**d.** $2 - (-9)$
↓	↓	↓	↓
____ ____ = ______	____ ____ = ______	____ ____ = ______	____ ____ = ______

4. Simplify these expressions by removing the parentheses.

a. $5x - (-y) =$	**b.** $12 + (-8t) =$	**c.** $-19a + (-b) =$

Step 1 2 3

5. **a.** How is this pattern growing?

   b. Write a formula for the number of snowflakes in step n.

   c. In which step will there be 194 snowflakes?
   Write an equation and solve it.

Skills Review 44

1. Solve.

a. $-3(x + 7) \quad = \quad 6x$	b. $-8(y - 4) \quad = \quad 5(y + 4)$

2. **a.** Which expression below is for the area of a rectangle? Which one is for its perimeter?

$10a + 2b \qquad 5a \cdot b$

b. Sketch the rectangle, marking its side lengths.

3. Multiply.

a. $7 \cdot (-3) =$ ________	**b.** $4 \cdot (-6) =$ ________	**c.** $2 \cdot (-7) \cdot 2 =$ ________
$-9 \cdot (-5) =$ ________	$-9 \cdot 8 =$ ________	$(-1) \cdot (-4) \cdot (-5) =$ ________

4. Write an equation for each problem, and solve it. You can also solve the problems using some other method just to check that you get the same answer.

a. Seven-twelfths of a number is 42. What is the number?	**b.** Six-ninths of a number is 7.2. What is the number?

Skills Review 45

1. Divide and simplify.

a. $-3 \div (-12)$	**b.** $18 \div (-24)$	**c.** $-16 \div (-20)$

2. **a.** Solve the inequality $7x - 12 > 62$ in the set $\{5, 7, 9, 11, 13, 15\}$.

   b. Solve the inequality $4x + 13 \le 53$ in the set $\{4, 6, 8, 10, 12\}$.

3. Solve. Simplify the one side first.

a. $3x = -5 + 14 + (-3)$	**b.** $6 \cdot (-8) = -4z$

4. Write using symbols, and simplify if possible.

   a. the opposite of the absolute value of 15

   b. the absolute value of negative 16

   c. the opposite of the sum 6 + 8

   d. the opposite of the difference 7 − 4

5. For each problem below, write a proportion and solve it. Remember to check that your answer is reasonable.

a. A car can travel 62.0 miles on 4.6 gallons of gasoline. How far can it go with 9.4 gallons of gasoline?	**b.** In order to plant a new lawn, you need to plant 3 lb of grass seed per 1,000 ft^2. How many pounds of seed do you need to plant a 60 ft by 42 ft lawn?

Skills Review 46

1. Find the unit rate.

a. \$144 for 3 coats

b. \$8 for 40 note cards

c. \$2.63 for ½ pound

2. Multiply. Then use the same numbers to write an equivalent division equation.

a. $-4 \cdot (-6) =$ ______ ______ ÷ ______ = ______	**b.** $7 \cdot (-3) =$ ______ ______ ÷ ______ = ______	**c.** $-20 \cdot 9 =$ ______ ______ ÷ ______ = ______

3. Convert the given times into hours in decimal format. Round your answers to three decimals, if necessary.

a. 28 minutes	**b.** 51 minutes

4. Write the numbers with scientific notation.

a. 16,000

b. 307,000

c. 42,900

d. 9,108,000

5. Calculate the value of these expressions both with a calculator and using pencil and paper.

a. $\dfrac{9}{-2\frac{1}{4}}$	**b.** $\dfrac{-\frac{4}{5}}{6.4}$	**c.** $\dfrac{3.25}{\frac{2}{5}}$

6. Add.

a. $(-6) + 9 + 2 + (-8) =$ ______

b. $-5 + (-5) + 14 + (-3) =$ ______

Skills Review 47

1. Write an inequality for the problem, and solve it.

Allison painted 12 paintings to sell at an art fair. How much should she charge for each one if she wants to have at least \$400 left after she pays the \$45 vendor fee?

2. Write a single mathematical expression ("number sentence") for each situation. Don't write just the answer.

a. You buy n puzzles for \$25 each and m books for \$12 each. Write an expression for the total cost. *cost* =	**b.** The price of a \$30 blanket is discounted by 1/5. Write an expression for the current price. *price* =

3. Fill in the table of values and determine whether the two variables are in direct variation. Then plot the equation.

$y = x - 2$

x	−4	−3	−2	−1	0	1	2	3
y								

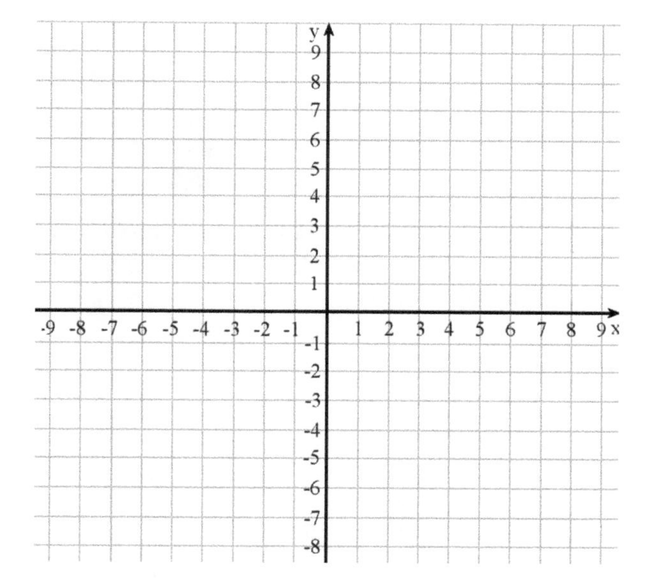

4. Solve.

a. $\frac{x}{4} = -6.39$	**b.** $\frac{s}{-0.9} = -0.7$

Skills Review 48

1. **a.** Does (−3, 2) fulfill the equation $y = 3x - 9$?

   b. Is (5, −2) a solution to the equation $y - 6 = -x$?

2. Determine the slope of each line from the table or from its graph.

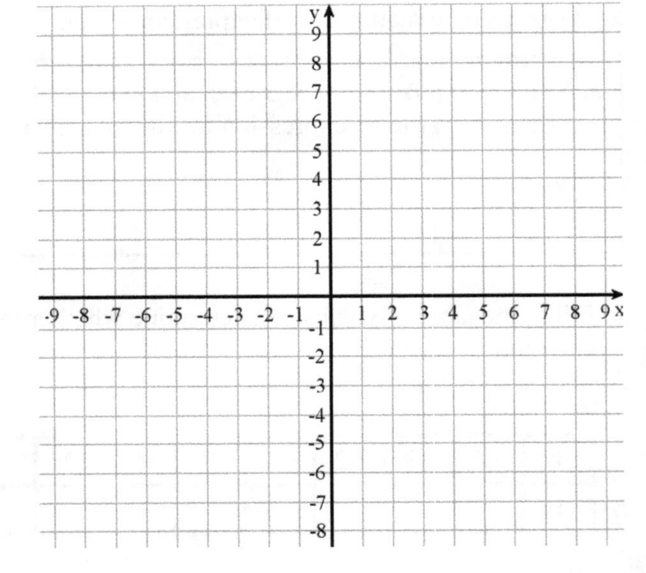

a.

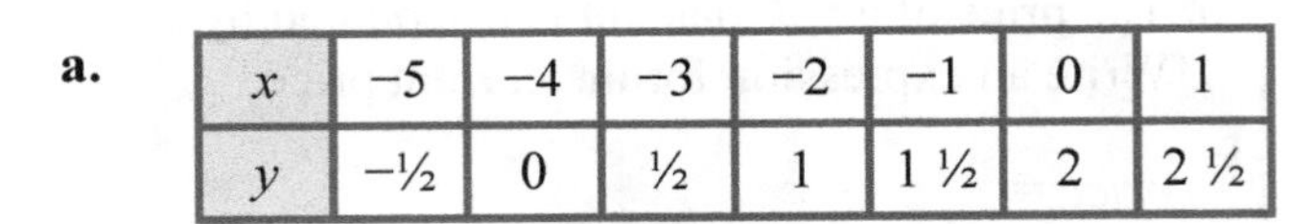

x	−5	−4	−3	−2	−1	0	1
y	−½	0	½	1	1 ½	2	2 ½

b.

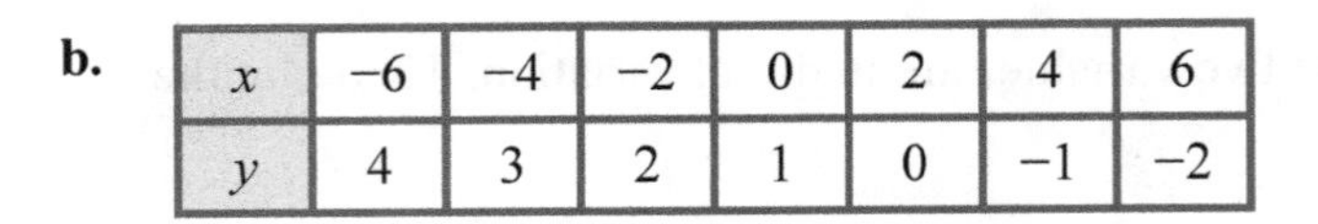

x	−6	−4	−2	0	2	4	6
y	4	3	2	1	0	−1	−2

3. The area of a square is 64 cm^2. The square is shrunk using the scale factor 3/4. What is the area of the resulting square?

4. Solve. Check your solutions (as always!).

a. $\dfrac{x+7}{4} = 18$	**b.** $\dfrac{x-3}{8} = -5$

Skills Review 49

1. Solve.

a. $38 + (-45) =$	b. $70 + (-26) =$	c. $-59 + 30 =$

2.

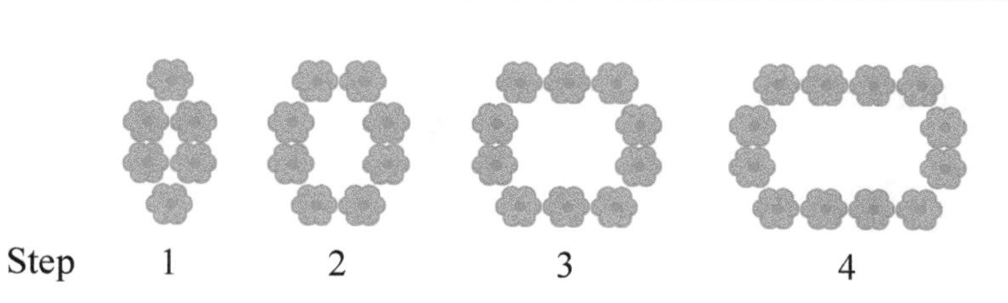

Step 1 2 3 4 5 6

a. Draw the next steps.

b. How do you see this pattern growing?

c. How many flowers will there be in step 39?

d. What about in step n?

3. A bus travels at a constant speed of 90 km/h from Chicago to New York City, a distance of approximately 1,270 km.

a. Write an equation relating the distance (d) it has traveled and the time (t) that has passed.

b. Plot your equation.

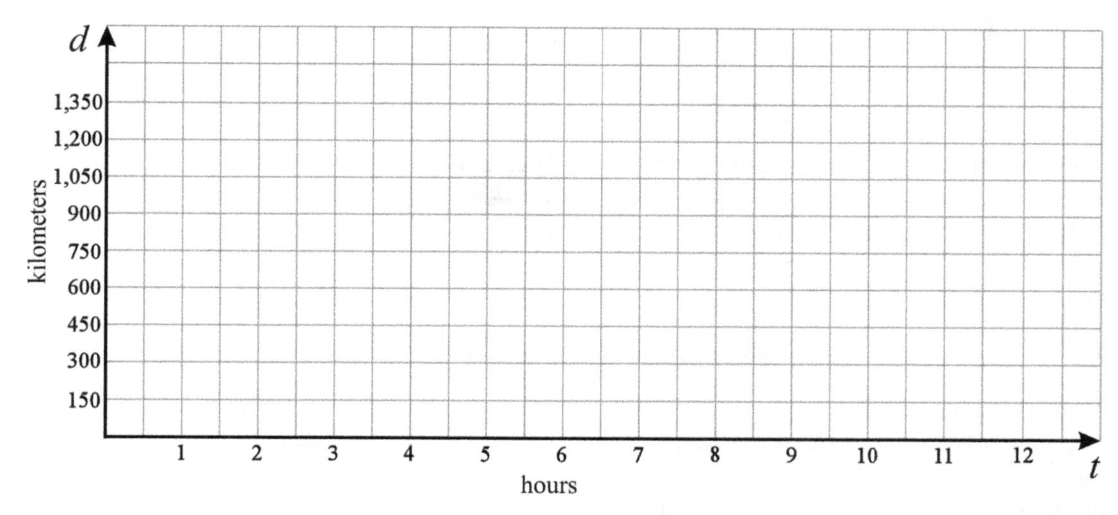

c. How far will the bus travel in 3 h 15 minutes?

d. At what time will the bus reach New York City if it left Chicago at 6:45 am?

Skills Review 50

1. A room measures 5 inches by 6 inches on a plan with a scale of 1 in : 2 ft.

   a. What would the room measure on the plan if it was drawn to the scale of 1 in : 3 ft?

   b. What would the room measure on the plan if it was drawn to the scale of 1 in : 5 ft?

2. Solve. Check your solutions (as always!)

a.  $2x - 76 = -4 - 7x$	**b.** $-4y - 9 = 63 + 2y$

3. Simplify these rates. Don't forget to write the units.

a. 540 km per 9 hours	**b.** 6.5 inches : 2.5 minutes

4. Change each addition into a subtraction or vice versa. Then solve whichever is easier.

a. $-7 + (-10)$ ↓ ____ ____	**b.** $12 - (-9)$ ↓ ____ ____	**c.** $2 + (-6)$ ↓ ____ ____	**d.** $5 + 8$ ↓ ____ ____

5. Which decimal is greater?

a. Which is more, $0.28\overline{0}$ or 0.28? How much more?	**b.** Which is more, $0.\overline{75}$ or 0.75? How much more?

Skills Review 51

1. Write as percentages, fractions, and decimals.

a. 39.4% = $\frac{\quad}{\quad}$ = ______	**b.** ______% = $\frac{216}{1000}$ = ______	**c.** ______% = $\frac{\quad}{\quad}$ = 0.702

2. A map has a scale ratio of 1:20,000. Fill in the table.

on map (cm)	in reality (cm)	in reality (m)	in reality (km)
1 cm	20,000		
4 cm			
6.5 cm			
0.9 cm			
12.7 cm			

3. Multiply.

a. $-6 \cdot 8 =$ ______ $-4 \cdot (-7) =$ ______	**b.** $(-2) \cdot (-9) =$ ______ $5 \cdot (-12) =$ ______	**c.** $(-10) \cdot 30 \cdot (-2) =$ ______ $-3 \cdot (-50) \cdot 4 =$ ______

4. Simplify the expressions.

a. $y \cdot y \cdot x \cdot y \cdot 9 \cdot y \cdot x$	**b.** $d + a + a + d + a + d$	**c.** $z + z + 5 + z$

5. Solve. Check your solutions.

a. $5 - (-12) = x + 9$	**b.** $3 - 10 = 4 + w$

Skills Review 52

1. Solve.

a. $4(x+3) = 3(x-(-1))$	b. $6(y-9)+3 = 15$

2. Simplify the expressions.

a. $7s + 9s - s$	b. $p^2 + 5p^2 + 8p^2$	c. $11x^2 - 6x^2 + 2x^2$

3. Write an equation for the situation. Then solve it. Do not write the answer only.

Five farmers divided a large tract of land equally,
so that each one received 1,800 hectares of land.
How many hectares was the original tract of land?

4. Make up a situation from real life that could be described by the given inequality.

a. $p > 85$

b. $x \leq 40$

5. Mary earns $120 for nine hours of work. In how many hours will she earn $800?

Earnings							
Work Hours							

Chapter 7

Skills Review 53

1. Solve the following proportions by using cross-multiplication. Give your answers to the nearest hundredth.

a. $\frac{T}{30} = \frac{72}{8}$	b. $\frac{13}{152} = \frac{4}{M}$

2. Calculate the new, increased prices. Write the percentages as decimals and use multiplication.

a. Bicycle: Original price \$594.85, increase 8%.

New price = _________ · \$594.85 = _______________

b. Telescope: Original price \$769.90, increase 4.5%.

New price = _________ · \$769.90 = _______________

3. Write the unit rate as a complex fraction, and then simplify it.

a. 5 ¼ feet of string costs \$3.
b. It took Susan 3/4 of an hour to watch 2/3 of a documentary.

Skills Review 54

1. Use the formula $d = vt$ to solve the problems.

a. A helicopter flies at a constant speed of 95 mph. How long will it take it to fly 215 miles?	**b.** Beth leaves at 7:05 a.m. to drive 35 miles to work. If her average speed is 42 mph, when will she arrive at work?
$d = v\ t$ ↓ ↓ ↓	$d = v\ t$ ↓ ↓ ↓

2. The table shows the relationship between the number of people and the time it takes to decorate a large room for a wedding.

Number of people	1	2	3	4	5	6
Time (hours)	27	22	18	15.5	13	10.8

a. Are these two quantities in proportion?

b. If so, write an equation relating the two and state the constant of proportionality.

3. A stove is discounted by 34%, and now it costs \$349.80. Let p be its price before the discount.

a. Find the proportion on the right that matches the problem.

b. Solve the problem.

$349.80/p = 66/100$

$p/349.80 = 66/100$

$p/66 = 349.80$

4. Solve (without a calculator).

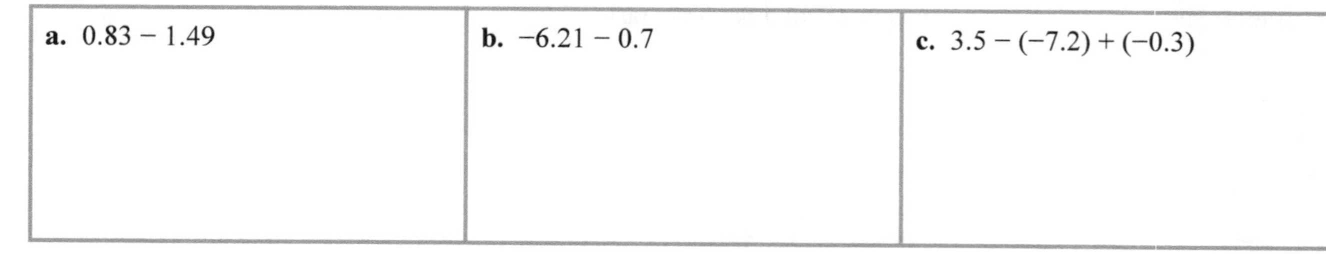

a. $0.83 - 1.49$	**b.** $-6.21 - 0.7$	**c.** $3.5 - (-7.2) + (-0.3)$

Chapter 7

Skills Review 55

1. Rewrite the numbers with the correct scientific notation.

a. $245 \cdot 10^6$

b. $0.627 \cdot 10^4$

2. Sketch a circle graph that shows...

a. 45%, 35%, and 20%	**b.** 55%, 20%, 1/8, and 1/8	**c.** 15%, 20%, 5%, and 60%
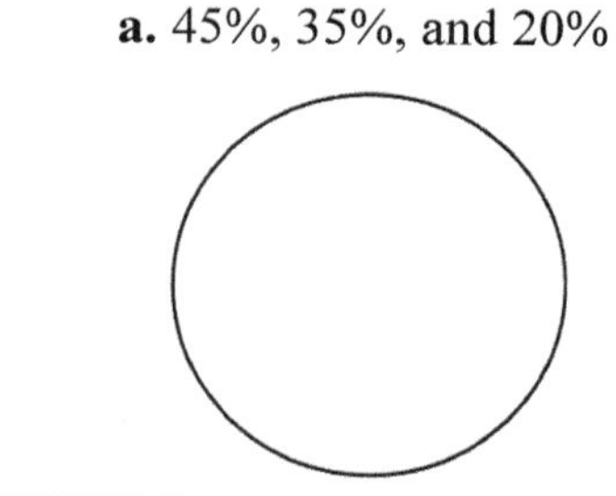	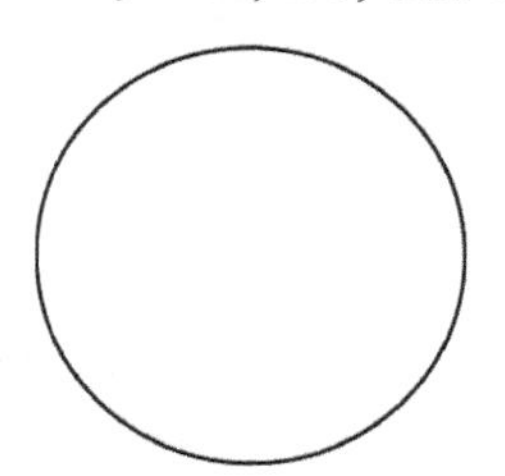	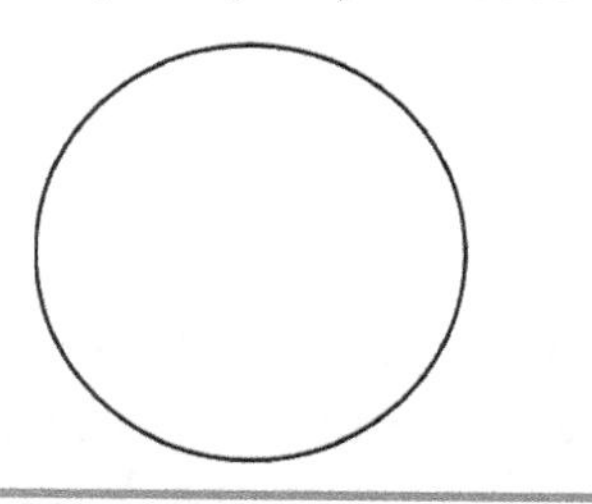

3. Complete the equations. There are many possible solutions. Find at least two different ones!

a.	**b.**
_____ + _____ + _____ = −5	_____ + _____ + _____ = 1
_____ + _____ + _____ = −5	_____ + _____ + _____ = 1

4. Solve the equations.

a. $x + \frac{1}{4} = \frac{11}{20}$	**b.** $x - \frac{2}{9} = \frac{1}{6}$

5. Divide and simplify.

a. $-2 \div (-6)$	**b.** $18 \div (-12)$	**c.** $-20 \div (-35)$

Skills Review 56

1. *Write an inequality for the problem, and solve it. Also, explain the solution set in words.*

You have a coupon that gives you a $15 discount on your total purchase at Fawna's Fashion. You can afford to spend at most $80. What is the maximum number of blouses that cost $13.45 each that you can buy?

2. **a.** At a fruit market, there are three times as many oranges as there are apples, and four times as many bananas as there are oranges. Let's denote the number of oranges with x. Write an expression for the amount of apples, in terms of x.

   b. Write an expression for the amount of bananas in terms of x.

3. One particular bottle of shampoo costs $9.60 for 30 fl oz and another costs $3.72 for 12 fl oz. Are the two rates equal? If not, which shampoo costs more per fluid ounce?

4. Continue the pattern.

$15 \div 3 =$ ______

$9 \div 3 =$ ______

$3 \div 3 =$ ______

$(-3) \div 3 =$ ______

______ $\div 3 =$ ______

______ $\div 3 =$ ______

______ $\div 3 =$ ______

______ $\div 3 =$ ______

5. Calculate the percentage of change.

A digital camera originally cost $350. It is discounted and now costs $308. What percentage is the discount?

$$\frac{\textit{difference}}{\textit{original}} =$$

Skills Review 57

1. Draw two lines with a slope of 3/5. They can be drawn anywhere on the grid; they do not have to go through any specific point.

   Check: Your lines should be parallel.

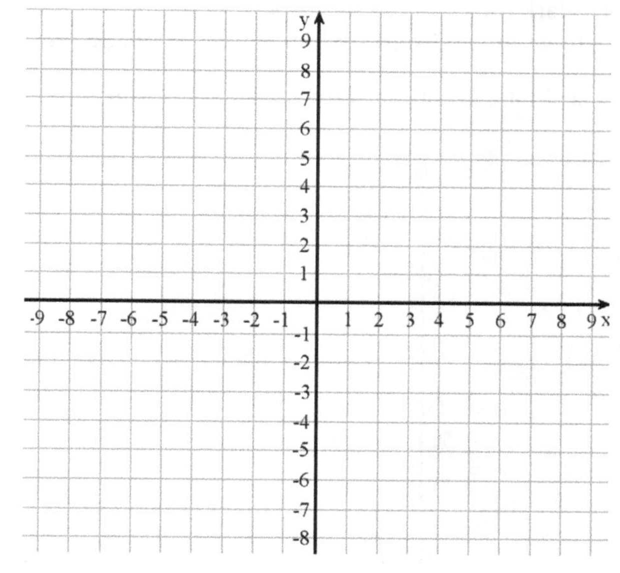

2. Draw any triangle on blank paper. Then draw another, bigger triangle using the similarity ratio 3:7. Remember that corresponding angles in the two triangles will be equal.

3. Solve *without* a calculator. Change decimals into fractions or treat fractions as divisions, whichever is easier.

a. $-\frac{1}{7} \cdot 0.8$	**b.** $\frac{1}{8} \cdot (-2.6)$	**c.** $0.4 \cdot \frac{3}{5}$

4. Solve the problem using an equation and also using some other strategy, such as a bar model or mental reasoning.

Four-ninths of a number is 5.62. What is the number?	
Equation:	Another way:

Skills Review 58

1. Mark runs at a constant speed of 5 m/s for eight seconds. He falls down and it takes him six seconds to get up and start running again. Then, he runs to the finish line (at 100 m) at a speed of 7 m/s.

   a. Plot a graph for the distance Mark runs.

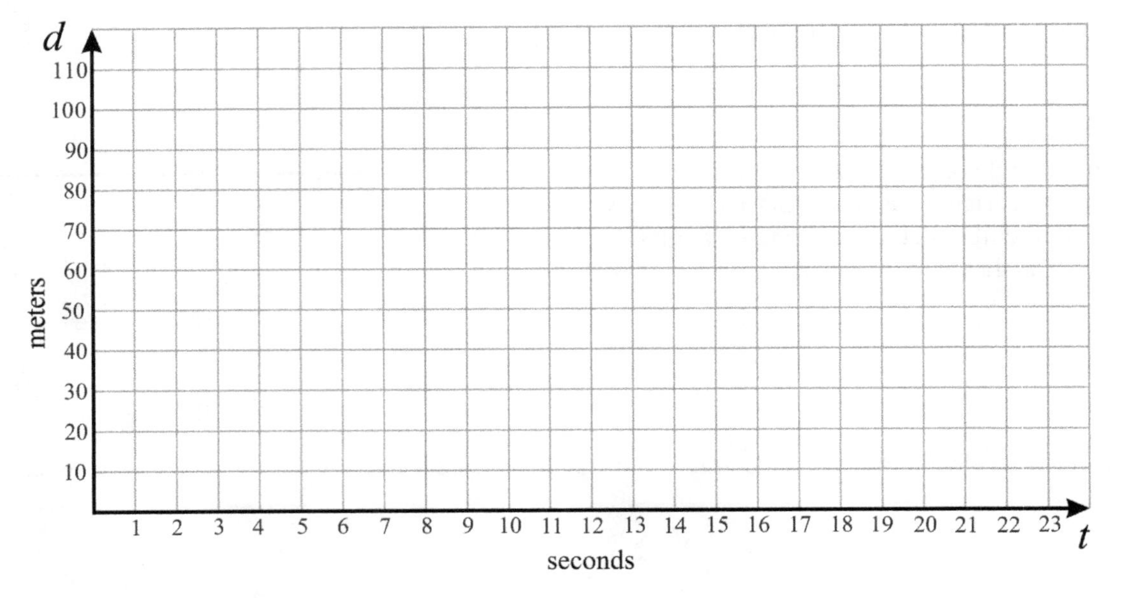

   b. How long does he take to run the race?

2. Find the value of the expressions with a calculator. Give your answer rounded to three decimal digits.

a. $\frac{4}{13} \cdot (-0.267)$	**b.** $8.51 \cdot \frac{7}{9} \cdot 0.3$	**c.** $0.49 \cdot \frac{26}{37} \cdot \frac{15}{10}$

3. The area of Oregon is about 96,000 square miles, and the area of North Dakota is about 69,000 square miles. Use their average area to calculate the relative percentage difference between their areas.

4. Solve.

a. $-8 = \frac{-6x}{9}$	**b.** $3 - 5 = \frac{s + 7}{4}$

Skills Review 59

1. Graph the equations as lines.

   a. $y = 2 - x$

   b. $y = (1/2)x + 1$

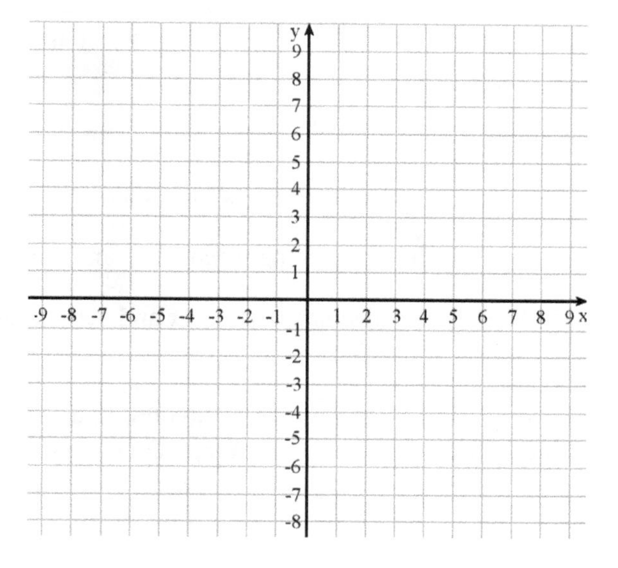

2. On a blank sheet of paper, draw a plan for a room that measures 2.8 m by 4 m. Put a 60 cm by 90 cm desk in the middle of the room. Use the scale 4 cm : 1 m.

3. Figure out the missing entries in the table without actually measuring any angles. Round the percentages to the nearest tenth of a percent. Recall that a full circle is 360°.

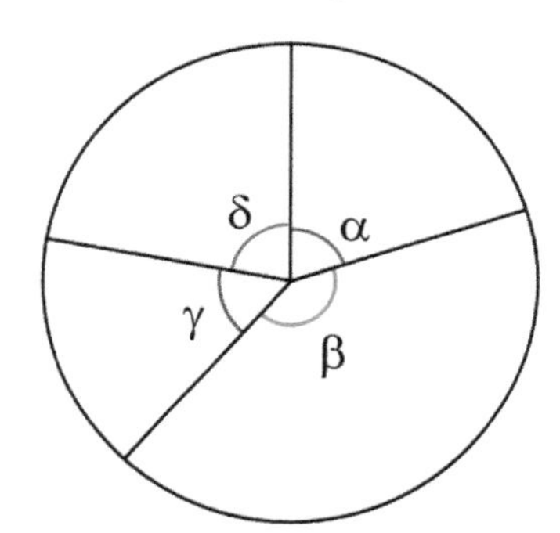

Angle	Degrees	Fraction	Percentage
α		1/5	
β	150°		
γ			
δ	80°		

4. Write the equivalent ratios.

a. 6 to 42 = 1 to ________	**b.** 4 : _____ = 12 : 96	**c.** 160 : 400 = 2 : _____	**d.** $\frac{7}{11} = \frac{\square}{143}$

5. Find the final price when the base price and sales tax rate are given. This is a mental math workout, so do not use a calculator!

a. Camera: \$70; 8% sales tax.	**b.** Sofa: \$300; 5% sales tax.	**c.** Game: \$40; 7% sales tax.
Tax to add: \$____________	Tax to add: \$____________	Tax to add: \$____________
Price after tax: \$__________	Price after tax: \$__________	Price after tax: \$__________

Skills Review 60

1. Form a fraction from the two given integers. Then convert it into a decimal.

a. 7 and 12	b. −3 and 8	c. 65 and −100

2. Answer. Use absolute values to calculate your answers.

   a. What is the distance between −183 and zero on a number line?

   b. What is the distance between x and zero on a number line?

3. A map has a scale ratio of 1:300,000. In miles, how long is a town that measures 3.2 inches on the map? Give your answer to the nearest mile.

4. Solve. Check your solutions.

a. $4x + 5x + 2 = 12x - 2x - 7$	b. $15y - 9y - 6 = 9y - y$

5. Calculate the interest and the total amount to be paid back on these investments.

   a. Principal \$8,000; interest rate 7%; time 1 year

   Interest: ______________________ Total to withdraw: ______________________

   b. Principal \$6,300; interest rate 5.4%; time 3 years

   Interest: ______________________ Total to withdraw: ______________________

Skills Review 61

1. Write an **expression** for the final price using a decimal for the percentage.

   a. Jacket: price \$45, discounted by 32%. New price = _______________

   b. A chainsaw: price p, discounted by 17%. New price = _______________

2. Use the distributive property to multiply.

a. $-0.7(w + 18) =$	**b.** $-0.2(9x - 13) =$	**c.** $-400(0.5x + 0.8) =$

3. **a.** Calculate the rate of physicians per 10,000 people in Nigeria, if the country is estimated to have 75,000 doctors and 205,700,000 people. Round your answer to one decimal.

   b. Sri Lanka has 6.8 physicians per 10,000 people. How many doctors would you expect to find in an area in Sri Lanka that has 430,000 residents?

4. Find the angle measure of the angle marked with a question mark.

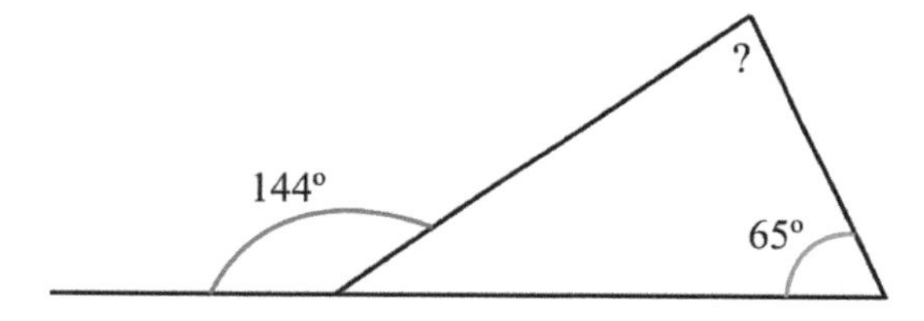

5. Solve.

a. $(-80) + 68 =$	**b.** $-30 + (-23) =$	**c.** $-91 + 45 =$

6. The table lists by variety how many bags of dried beans a supermarket sold. Draw a circle graph. You will need a protractor and a calculator. Round the percentages to the nearest tenth of a percent. Round the angles to the nearest degree.

(With the rounded numbers, your totals might not add up to exactly 100% or 360°, but they are close enough to make the circle graph.)

Variety	Amount sold	Percentage of total	Central Angle
fava	16		
kidney	29		
navy	22		
pinto	37		
TOTALS			

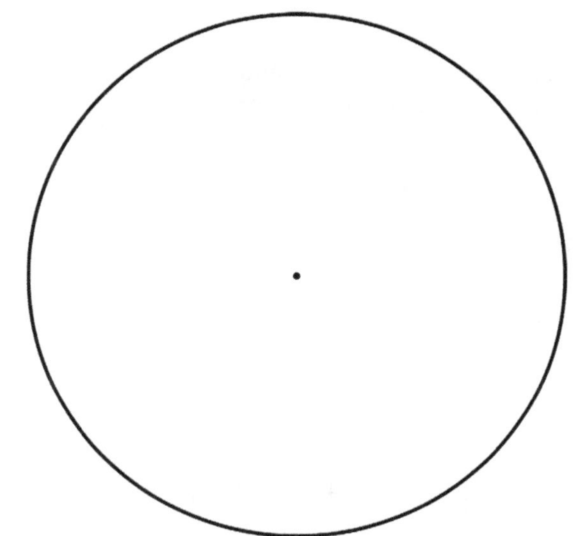

Skills Review 62

1. Draw a triangle with sides 3.9 cm, 5.6 cm, and 8.2 cm long.
   You will need a regular centimeter-ruler and a compass.

2. Think up a real-world context for each calculation.

   a. −40 ft − 10 ft

   b. \$30 − \$55

3. Solve. Check your solutions.

a. $-6 + r = -4 + (-9)$	**b.** $12 - (-5) = 6 + 8 + t$

4. A bromine ion has 35 protons and 36 electrons. The electric charge of the protons is $35e$, and the charge of the electrons is $-36e$. What is the total electric charge of this ion?

5. Solve the inequalities and plot their solution sets on a number line. Write appropriate multiples of ten under the bolded tick marks (for example, 30, 40, and 50).

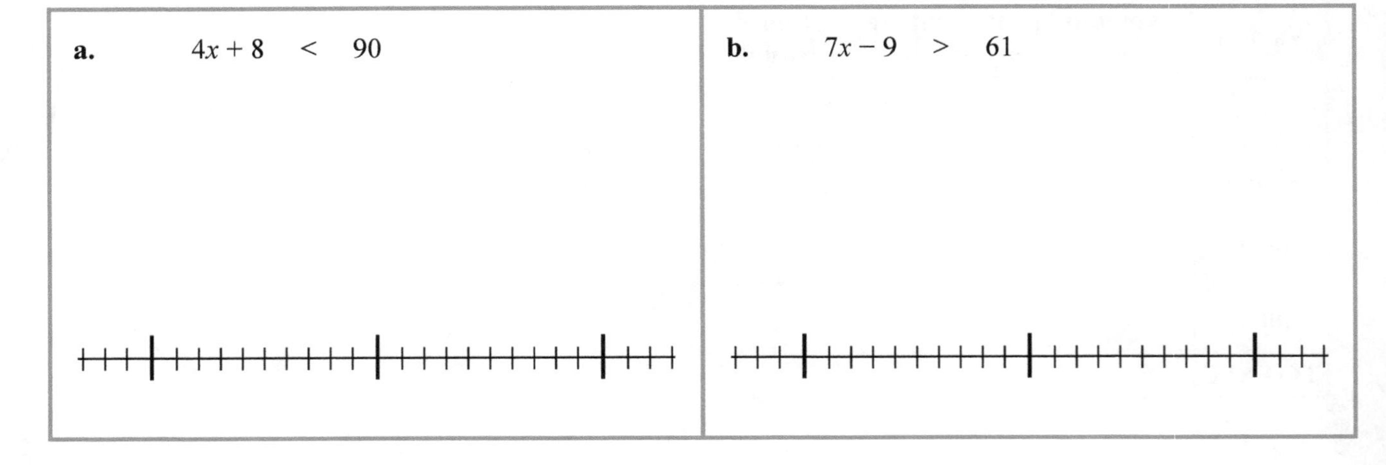

Chapter 8

Skills Review 63

1. Solve and compare the two problems.

a. A lamp used to cost \$50 but it was discounted by 20%. What is the new price?	**b.** A lamp used to cost \$50. Now it is discounted to \$35. What percentage was it discounted?

2. In a poll that interviewed 2,000 people about their favorite flower, 720 people said they liked roses.

   a. Simplify this ratio to lowest terms.

   b. Assuming the same ratio holds true in another group of 240 people, how many of those people can we expect to like roses?

3. Add (without a calculator).

a. $-0.9 + 0.4 + (-2.5) + (-5.7) + 3.1$	**b.** $-\$2.06 + (-\$6.20) + \$0.87 + \$0.93 + (-\$0.42)$

4. Does the information given define a unique triangle? If yes, say so, and draw the triangle. If not, prove that it doesn't by drawing at least two non-congruent triangles that satisfy the given conditions.

   a. A triangle with two 50-degree angles

   b. An isosceles triangle with two 5.5-cm sides and a 8-cm base

Skills Review 64

1. Find the missing factors.

a. 9 · ______ = −72	**b.** −6 · ______ = 90	**c.** 8 · ______ = −160

2. Seth is 135 cm tall, and Barry is 190 cm tall. Fill in the blanks.
(Round the answers to the nearest percentage point.)

Barry is 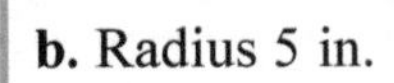 = ______% taller than Seth. Seth is 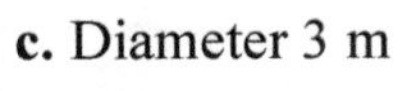 = ______% shorter than Barry.

3. Find the circumference of each circle. Use 22/7 for pi. Round the answers to one decimal.

a. Diameter 7 cm Circumference =	**b.** Radius 5 in. Circumference =	**c.** Diameter 3 m Circumference =

4. For each problem below, write a proportion and solve it. Carry the units through your calculation. Don't forget to check that your answer is reasonable.

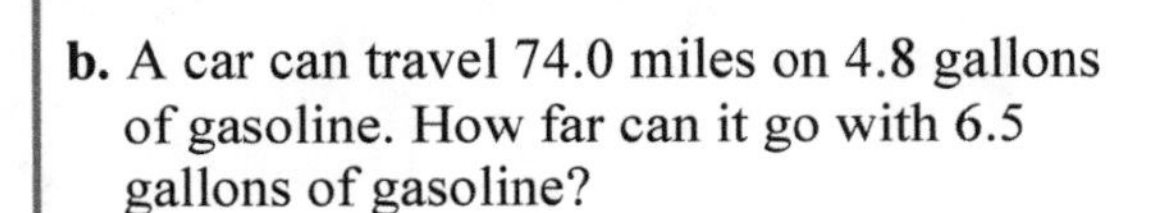

a. It costs $1,330 to pave 140 ft^2 of a driveway. How much does it cost to pave 260 ft^2?	**b.** A car can travel 74.0 miles on 4.8 gallons of gasoline. How far can it go with 6.5 gallons of gasoline?

5. Give these times in hours and minutes.

a. 2.638 hours	**b.** 18/40 hours

Skills Review 65

1. Solve. Check your solutions.

a. $-5 = -\frac{1}{7}x$	**b.** $-32 = \frac{1}{9}x$	**c.** $\frac{1}{11}x = -12 + 7$

2. **a.** Bonnie opened a savings account with $8,800 that paid an interest rate of 7.5%. In how many years will her account contain $15,400?

b. Eric deposited $8,000 into a savings account for two years. After that, the account had $8,400. What was the interest rate?

3. Determine whether the two quantities are in proportion. If so, find the unit rate, write an equation relating the two, and graph the equation.

x	−6	−5	−4	−3	−2	−1	0	1
y	9	7 ½	6	4 ½	3	1 ½	0	−1 ½

In proportion or not?

Unit rate:

Equation:

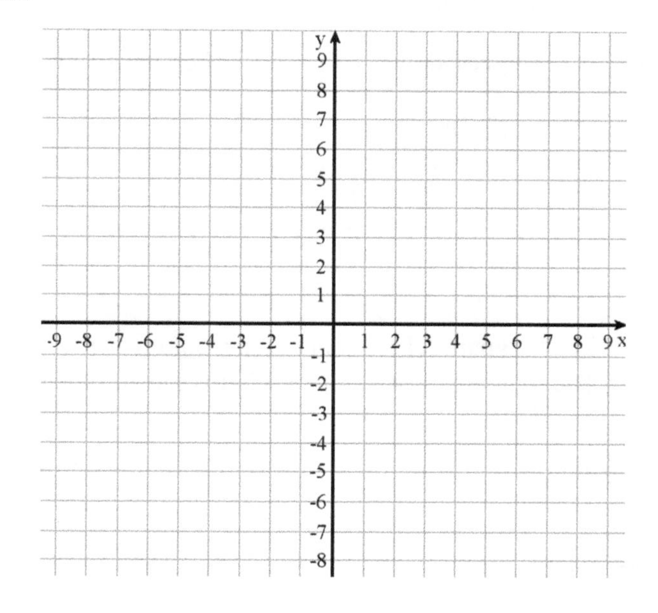

4. The diameter of a certain circle is 8 units. Which expression can you use to calculate the area of that circle?

a. $\pi \cdot 8$ **b.** $\pi \cdot 4^2$ **c.** $\pi \cdot 8^2$ **d.** $\pi \cdot 16$

Skills Review 66

1. Determine the slopes from the graphs. Remember that for a decreasing line, the change in the y-coordinates is negative, which makes the slope negative.

a. 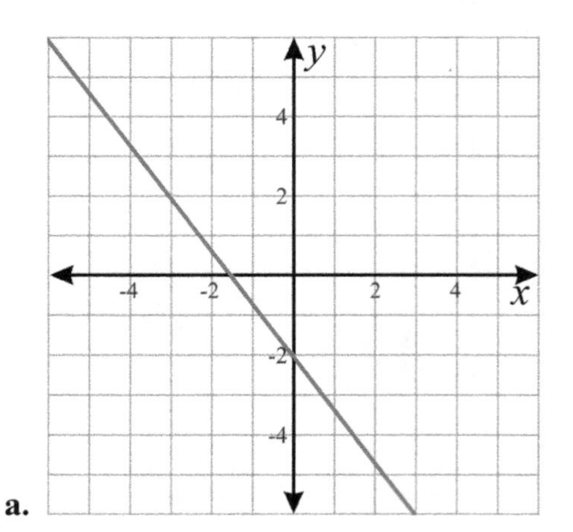

b. 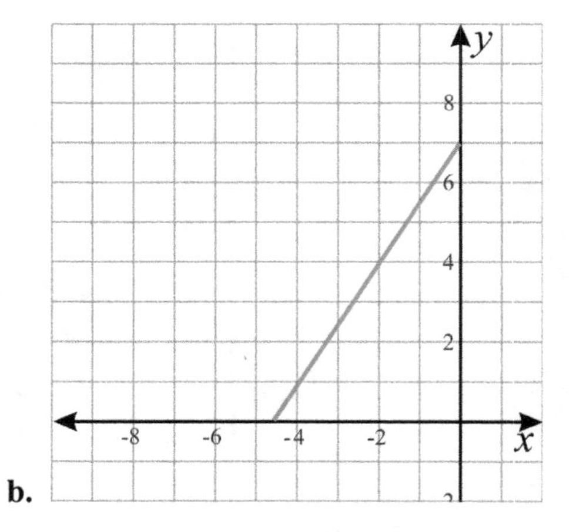

Write an inequality for the following problem, and solve it. Also explain the solution set in words.

2. Jeremy is driving a long distance to his grandma's house. Each day, he starts driving at 8 am and plans to quit traveling for the day at 6 pm. During the day, he stops for a 45-minute lunch break, and two 15-minute rest breaks. How many miles per hour will he have to drive, if he wants to travel at least 500 miles in a day?

3. Fill in the table.

Expression	the terms in it	coefficient(s)	Constants
$(3/7)s$			
$0.5x + 9.2y$			
$x^3y^8 + 6$			

4. The two parallel sides of a trapezoid measure 70 m and 95 m.
The total area is 4,125 m^2. What is the altitude of the trapezoid?

5. Divide and simplify.

a. $-49 \div 63$	**b.** $-94 \div (-15)$	**c.** $72 \div (-36)$

Skills Review 67

1. Add and subtract. First, change each subtraction into an addition.

a. $-11 - (-7) - (-3) =$	**b.** $8 - (-4) + (-9) =$

2. Mr. Jenkins flew his helicopter at a constant speed of 250 km/h from Atlanta to Jacksonville, a distance of approximately 560 km. After taking a two-hour break for lunch and to refuel, he then flew at the same speed from Jacksonville to Tampa, a distance of approximately 320 km.

a. How long did it take him to fly the first part of the trip?

b. And the second part?

c. At what time did the helicopter reach Tampa if it left Atlanta at 9:15 am?

d. Plot a graph over time for the distance Mr. Jenkins flew.

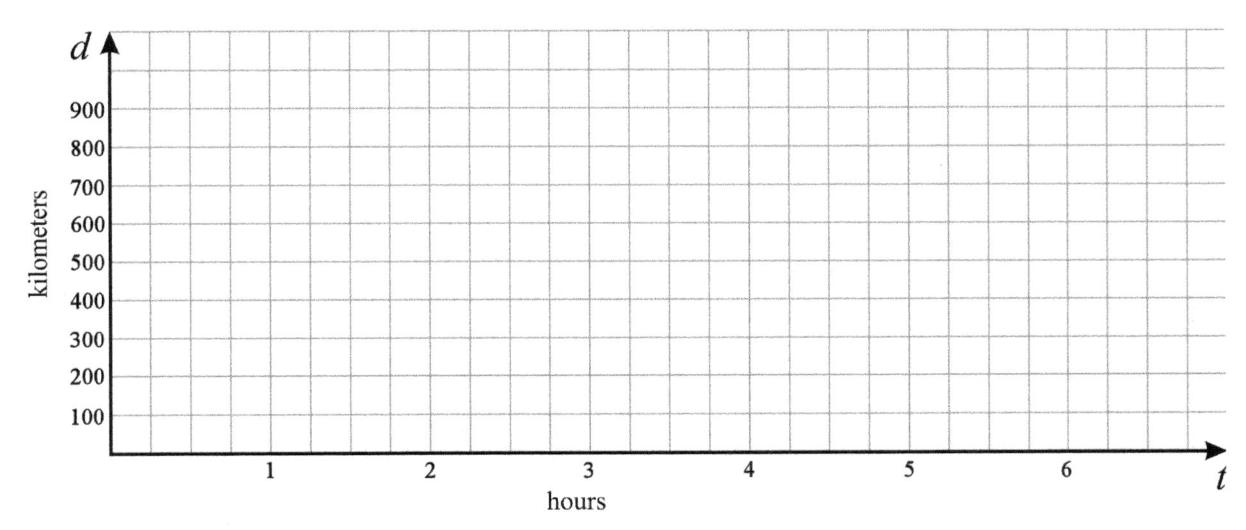

3. Calculate the value of these expressions both with a calculator and using pencil and paper.

a. $\dfrac{14}{-2\frac{1}{4}}$	b. $\dfrac{-\frac{5}{8}}{3.6}$	c. $\dfrac{5.75}{\frac{2}{10}}$

4. **a.** Draw a half-circle with a radius of 9 cm on blank paper. Cut it out and fold it into a cone.

b. Calculate the surface area of your cone (without the base circle) to the nearest square centimeter.

Skills Review 68

1. Multiply. Then use the same numbers to write an equivalent division equation.

a. $-9 \cdot (-4) =$ ______	**b.** $3 \cdot (-12) =$ ______	**c.** $-60 \cdot 7 =$ ______
______ ÷ _____ = ______	______ ÷ _____ = ______	______ ÷ _____ = ______

2. **a.** Does (−6, −20) fulfill the equation $y = 3x - 2$?

b. Is (4, −6) a solution to the equation $y - 2 = -x$?

3. A floor plan is drawn using the scale 4 cm : 1 m.

a. Calculate the dimensions in the plan for a bathroom that measures 3.7 m by 4.6 m in reality.

b. The living room measures 26.8 cm by 31.6 cm on the plan. What are its dimensions in reality?

4. Find the area in the given units. Then convert it to the other area unit.

Conversion Factors:
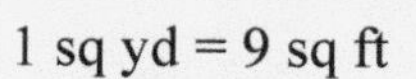
1 sq ft = 144 sq in
1 sq yd = 9 sq ft

a. A rectangle with 7.0 ft and 3.8 ft sides → A = ___________ sq ft Conversion: A = ___________ sq ft · $\frac{\text{sq in}}{\text{sq ft}}$ = ___________ sq in.
b. A rectangle with 18 ft and 12 ft sides → A = ___________ sq ft Conversion: A = ___________ sq ft · $\frac{\quad}{\quad}$ = ___________ sq yd.

5. Use a fraction line to write ratios of the given quantities. Then simplify the ratios.

a. 200 cm and 4.7 m	**b.** 5 ft 9 in and 2 ft 6 in

Skills Review 69

1. Calculate the measures of angles α, β, and γ.
   Hint: Look for vertical angles and for angles that form a straight line.

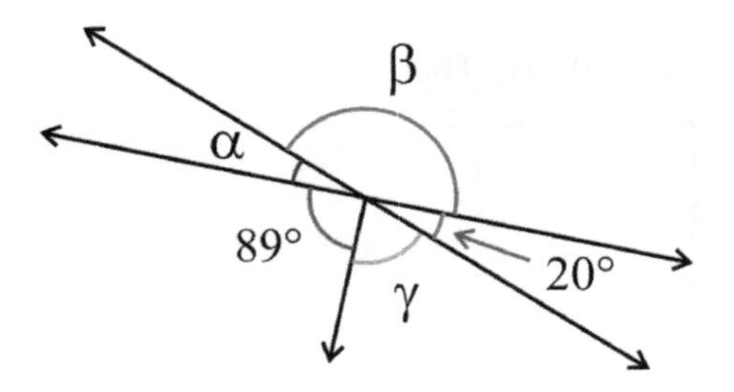

$\angle\alpha =$ ________ ° $\angle\beta =$ ________ °

$\angle\gamma =$ ________ °

2. Solve. Use a calculator. Round your answers to three decimal digits.

a. $\dfrac{c}{295.7} = -12.6$	**b.** $-0.83x = -2.09$

3. A town is spread out over a rectangular area with 2.7 km and 1.9 km sides. Calculate its area in square kilometers *and* in hectares.

4. Paula can ride her bicycle 6 miles in 50 minutes. At the same constant speed, how long will she take to go 27 miles? Fill in the equivalent rates below.

$$\frac{6 \text{ miles}}{50 \text{ minutes}} = \frac{3 \text{ miles}}{\square \text{ minutes}} = \frac{27 \text{ miles}}{\square \text{ minutes}}$$

5. Use the distributive property to multiply.

a. $-0.7(w + 24) =$	**b.** $0.5(8x - 60) =$	**c.** $-300(0.2x + 0.9) =$

6. Fill in the table with the rest of the numbers written in the indicated ways.

Scientific Notation	(in-between calculation)	Common notation
$6.45 \cdot 10^9$		
$4.212 \cdot 10^6$		
$7.3936 \cdot 10^2$		

Skills Review 70

1. Solve with a calculator. Round your answer to three decimal digits. Check that your answer makes sense by estimating with mental math.

a. $\frac{3}{15} \cdot (-2.074)$	**b.** $8.62 \cdot \left(-1\frac{5}{9}\right)$	**c.** 22% of $4\frac{3}{8}$

2. Find what is missing from the sums.

a. $9x + 4 +$ ______________ $= 3x + 6$	**b.** $7b - 3 +$ ______________ $= 4b + 8$
c. $-5z +$ ______________ $= 2 - 8z$	**d.** $-6f + 11 +$ ______________ $= -f - 9$

3. **a.** Martha needs 4 ½ cups of oats to make 5 dozen cookies. Find the unit rate for one cookie, as a fraction.

   b. It costs $30.10 for 3 ½ pounds of almonds. Find the unit rate for one pound.

4. The area of a square is 81 cm^2. The square is shrunk using the scale factor 2/3. What is the area of the resulting square?

5. Solve.

a. $\frac{3}{5}x = 480$	**b.** $\frac{5}{8}y = \$26.74$

Skills Review 71

1. The distance from Ann's home to her aunt's home is 67.39 miles, according to an online distance calculator.

   How long would this distance be, in inches, on a map with a scale of 1:350,000?

   How about on a map with a scale of 1:400,000?

2. Solve.

a. $\frac{x}{4} - (-14) = -9 \cdot 6$	**b.** $5 - 9p = -1.3$

3. Draw an isosceles triangle with a 74° top angle and two 4.8-cm sides on blank paper. What is the measure of the base angles?

4. Describe the cross section formed by the intersection of the plane and the solid.

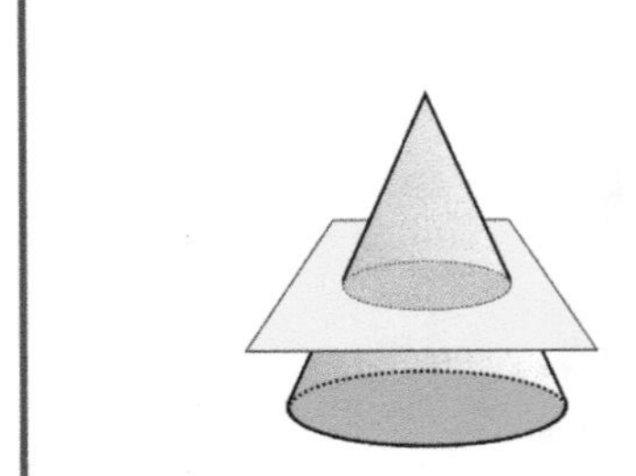	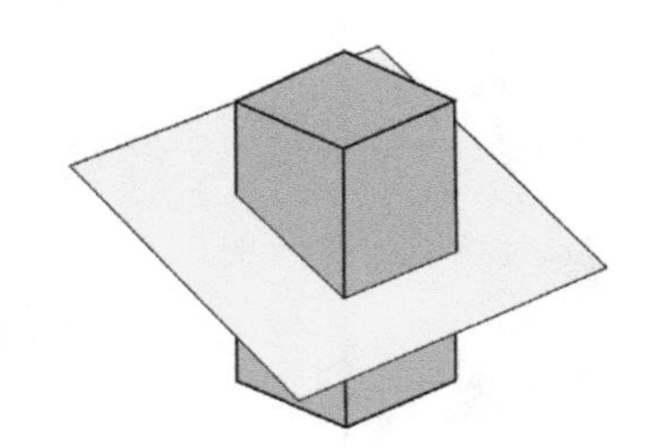	
a. The cross section is a ______________.	**b.** The cross section is a ______________.	**c.** The cross section is a ______________.

Skills Review 72

1. **a.** The base price of a purse is \$35.99. It is first discounted by 22% and then an 8% sales tax is added. What is the final price of the purse?

   b. The base price of a pair of boots is \$157.95. They are first discounted by 15% and then a 6% sales tax is added. What is the final price of the boots?

2. **a.** The table lists three sets of lengths. If these are used as lengths of sides for a triangle, one of them does not make a triangle. Which one? (Try to draw the triangles on a blank paper.)

7 cm, 9 cm, 12 cm	5 cm, 10 cm, 8 cm	11 cm, 4 cm, 6 cm

   b. Change one of the lengths in the set that didn't make a triangle so that the three lengths will form a triangle.

3. Martin works as an electrician. He gets paid \$224 for an 8-hour workday.

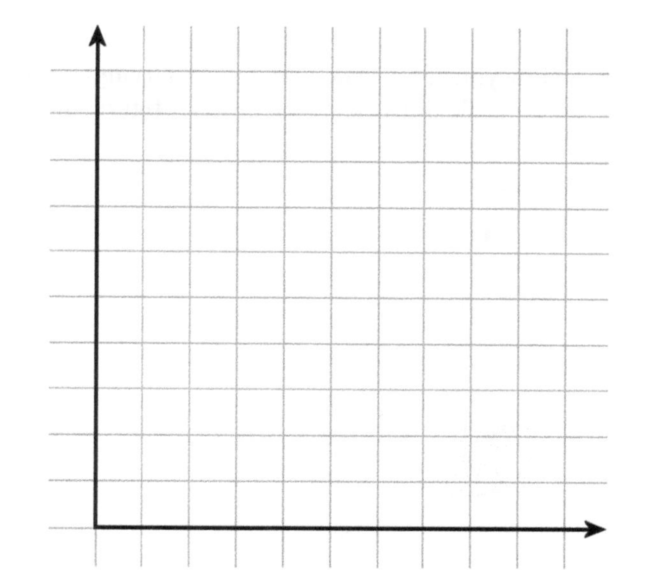

   a. Write an equation that relates Martin's pay to the time (in hours) that he has worked.

   b. Plot the equation on the grid. Choose the scaling on the axes so that you can fit the point that corresponds to *time* = 15 hours onto the grid.

   c. Plot the point that corresponds to the unit rate.

4. Solve the inequalities.

a. $4x - 8 \quad > \quad -7$	**b.** $12 + 5x \quad \leq \quad 50$

5. A cylinder-shaped barrel has a height of 27.5 inches and an inside diameter of 18.25 inches.

   a. Calculate its volume in cubic inches, to the nearest cubic inch. Use a value for Pi that is accurate to at least four decimal digits.

   b. Convert the volume to gallons, to the hundredth of a gallon.

1 gallon = 231 in^3

Skills Review 73

1. Determine the slope of each line from the table or from its graph.

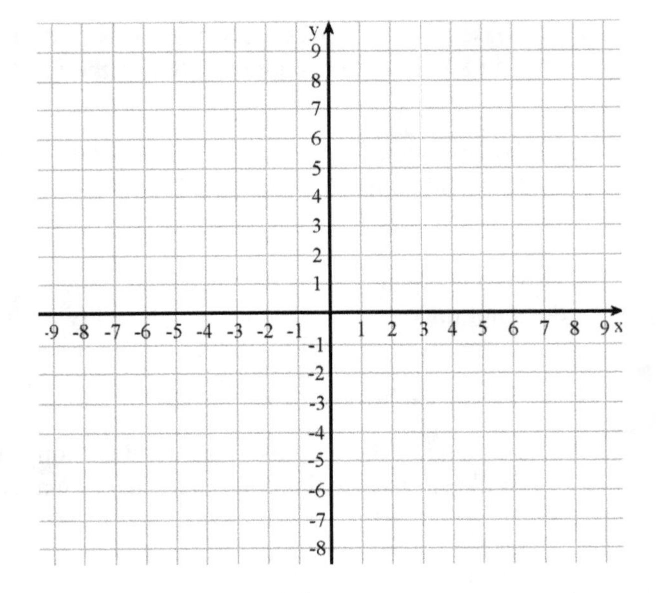

a.

x	−7	−5	−3	−1	1	3	5
y	−5 ½	−5	−4 ½	−4	−3 ½	−3	−2 ½

b.

x	9	8	7	6	5	4	3
y	−4	−3	−2	−1	0	1	2

2. Below you see a scale drawing of a triangle, drawn at the scale 1 cm = 25 cm. Make a new scale drawing of the original figure, this time using the scale 1 : 15.

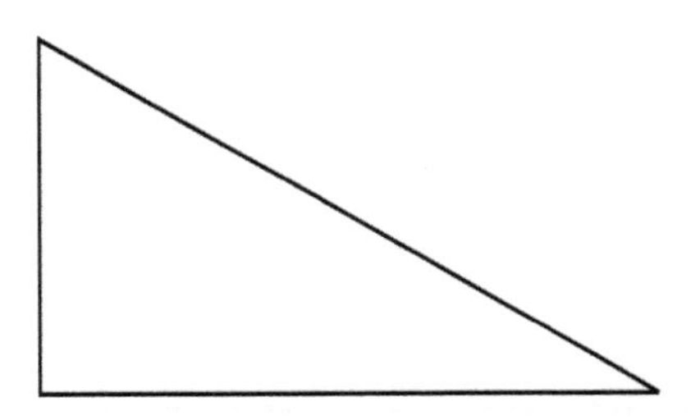

Scale 1 cm = 25 cm

3. Use a calculator to find these square roots. Round your answers to four decimal digits.

a. $\sqrt{10}$	**b.** $\sqrt{19}$	**c.** $\sqrt{27.43}$

4. Write and solve a percent proportion (according to the data below) in the form $\frac{part}{total} = \frac{percent}{100}$.

a. How much is 62% of 7,300 km?

b. Forty-three percent of a number is 8.17. What is the number?

Skills Review 74

1. Solve. Round the answers to three decimals. Check your solutions. You can use a calculator.

a. $a^2 + 4^2 = 9^2$	b. $36^2 + x^2 = 61^2$

2. Find the circumference of these circles. Use $\pi \approx 3.14$. Give your answer to the same decimal accuracy as the dimension given in the problem.

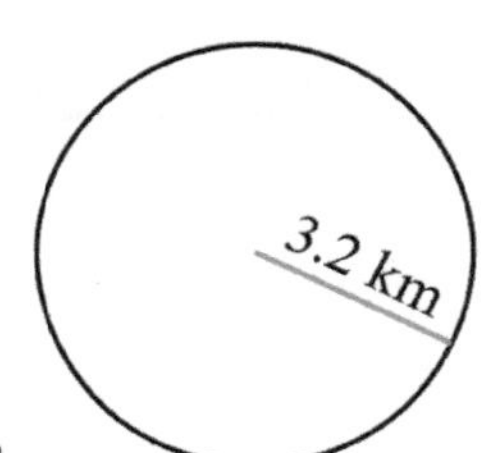

a.

C = ____________________

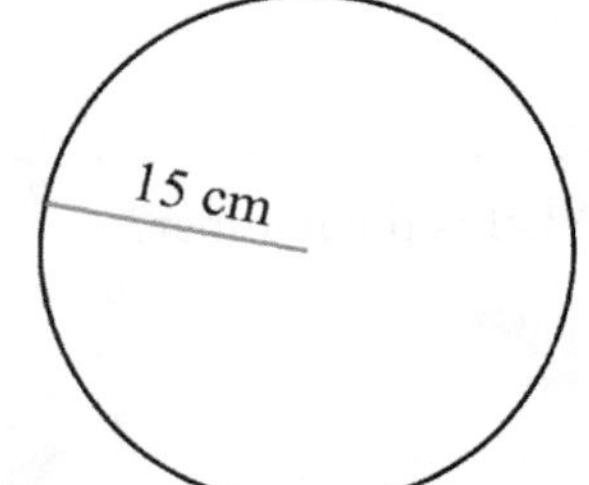

b.

C = ____________________

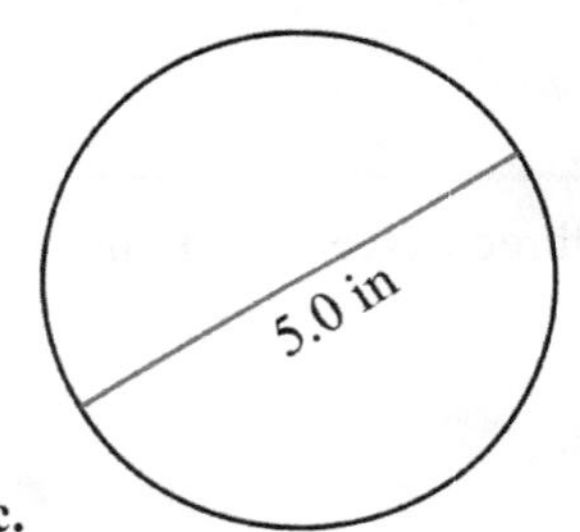

c.

C = ____________________

3. The Smith family spent $365 on groceries during the first week in April, and $412 the next week. The Daniels family spent $319 on groceries during the first week in April, and $345 on groceries the next week. Which family's grocery bill increased by the greatest percentage on the second week?

4. Evaluate the expressions. (Give your answer as a fraction or mixed number, not as a decimal.)

a. $\dfrac{x^3}{x+2}$, when $x = 5$	b. $\dfrac{x+3}{x-3}$, when $x = 12$

Skills Review 75

1. For each set of lengths, determine whether they form a right triangle using the Pythagorean Theorem. Notice carefully which length is the hypotenuse.

   a. 16, 12, 20

   b. 8, 11, 5

2. Seacoast City has 453,295 inhabitants and Snowflake City has 387,270 inhabitants. How many percent more inhabitants does Seacoast City have than Snowflake City?

3. Mike rides his motorcycle at a constant speed. The equation $d = 80t$ tells us the distance, in kilometers, that he travels in t hours.

   a. What is Mike's speed, in kilometers per hour?

   b. Plot the equation $d = 80t$.

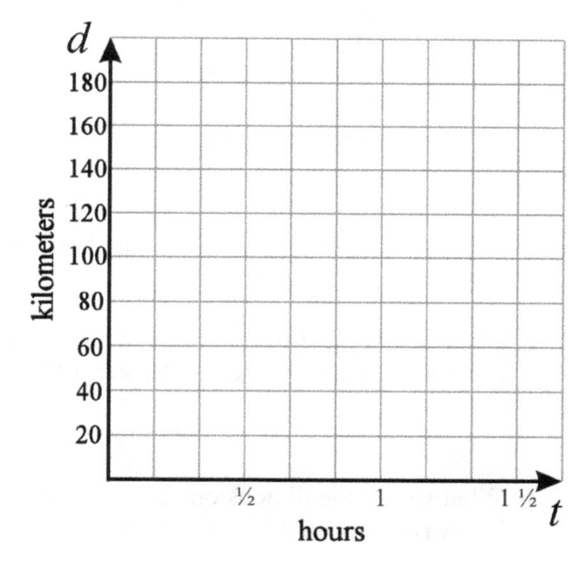

4. Find the areas of these circles.

a. A circle with a *diameter* of 63.0 cm. Round the answer to the nearest ten square centimeters (to 3 significant digits). Area =	**b.** A circle with a radius of 14 ft 9 in. Round the answer to the nearest hundred square inches (to 3 significant digits). Area =

Skills Review 76

1. The area of a square is 225 m^2. How long is the diagonal of the square (to the hundredth of a meter)?

2. Which equation matches the plot on the right?

$y = (1/4)x + 2$

$y = (1/4)x$

$y = (1/4)x - 2$

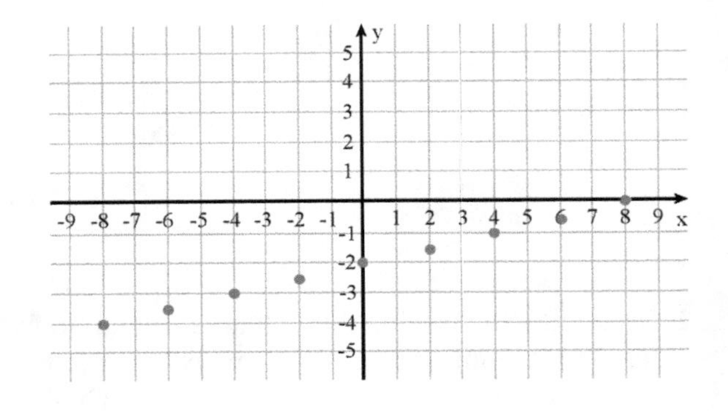

3. Find the area of the trapezoid.

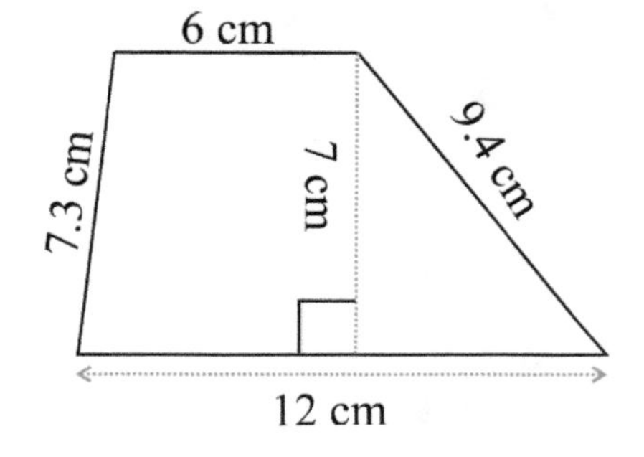

4. A room measures 6 ½ inches by 5 inches in a plan with a scale of 1 in : 2 ft.

a. What would the dimensions of the room be in the plan, if it was drawn to the scale of 1 in : 5 ft?

b. What would the dimensions of the room be in the plan, if it was drawn to the scale of 1 in : 3 ft?

5. Think of fractions. Estimate how many percent the sectors of the circle graphs represent.

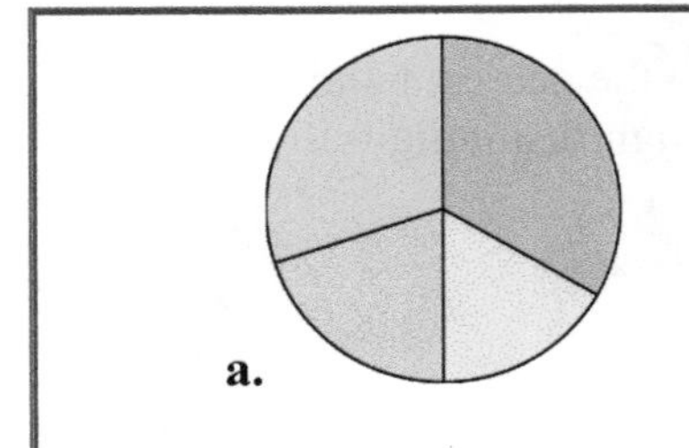

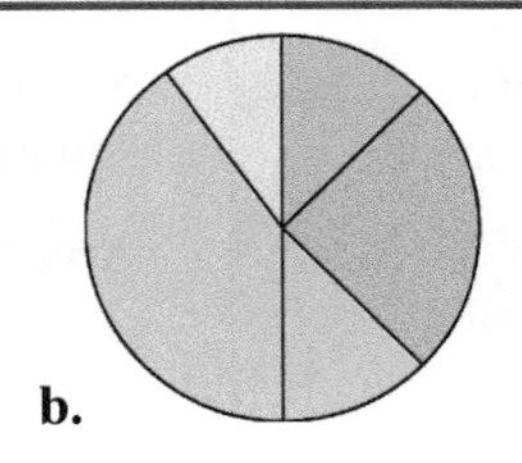

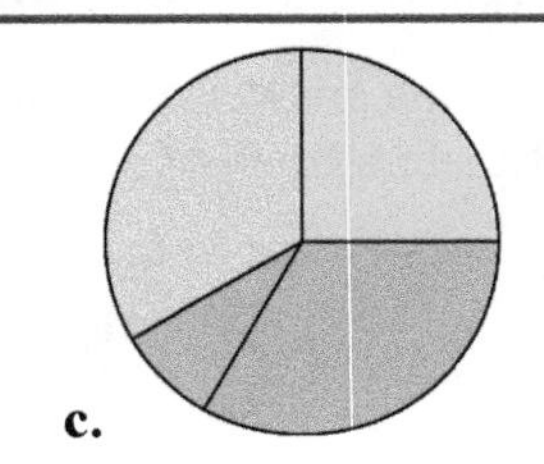

Skills Review 77

1. Calculate. Give your answers to three decimal digits.

a. $\sqrt{7 + 19}$	b. $\sqrt{13 \cdot 13}$	c. $\sqrt{3 \cdot (52 - 8)}$

2. Name the solid that can be built from each net.

a. 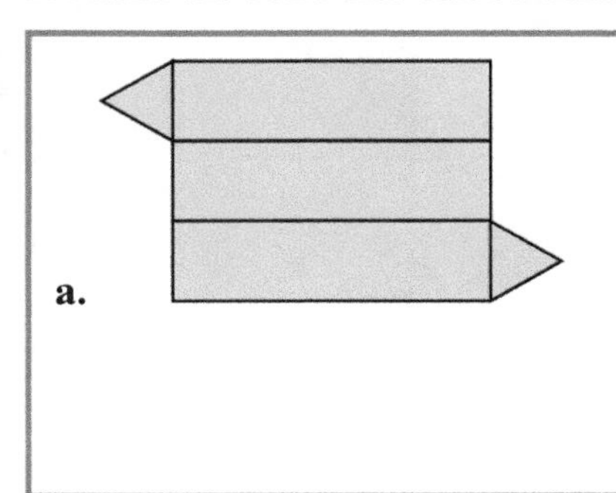	b. 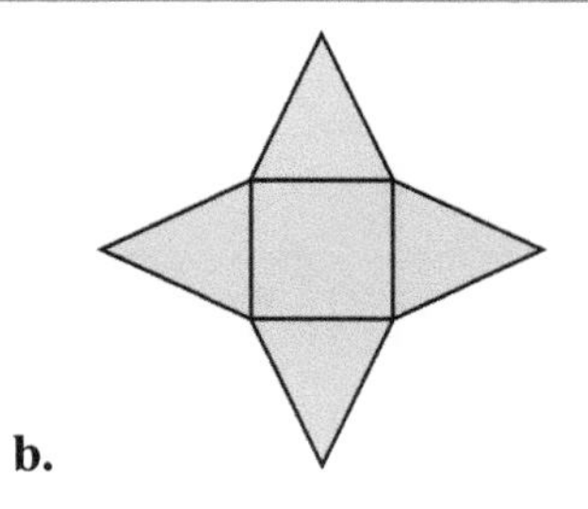	c. 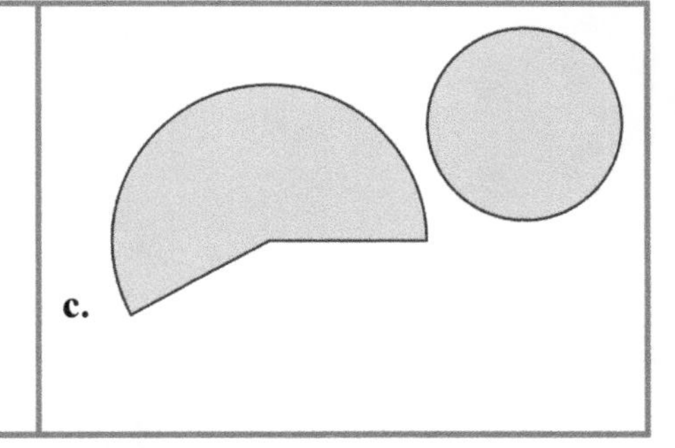

3. Find the surface area of this building (not including the bottom face).

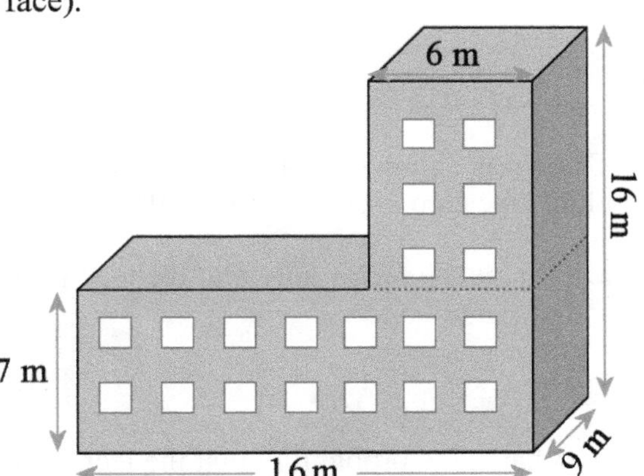

4. Suppose you choose one letter randomly from the word "THANKSGIVING."

a. List all the possible outcomes for this event.

Now find the probabilities of these events:

b. P(K)

c. P(I or G)

d. Make up an event for this situation that is likely to occur, yet not a sure event, and calculate its probability.

Skills Review 78

1. Solve. Remember, there will be two solutions. Check your solutions.

a. $4x^2 = 3{,}600$	**b.** $y^2 + 75 = 700$

2. The dotplot shows the age distribution of a teens' book club. One teen is chosen randomly from the group.

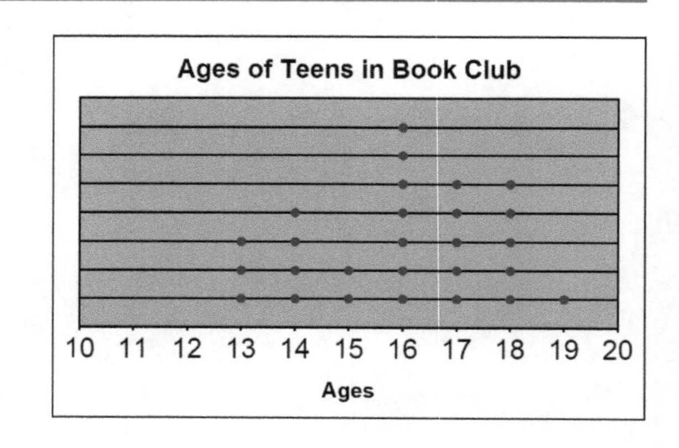

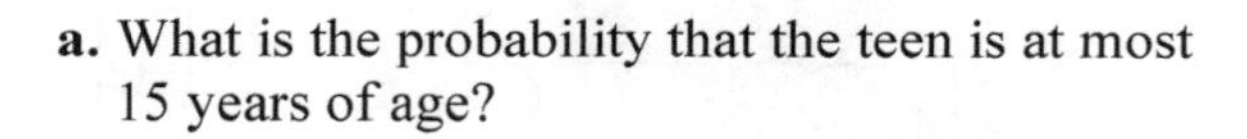

a. What is the probability that the teen is at most 15 years of age?

b. What is the probability that the teen is at least 17 years of age?

3. A rectangle measures 12 inches by 43 inches.

a. Calculate its area in square inches.

b. Convert the area into square feet, to the nearest hundredth of a square foot. (Remember, 1 ft^2 = 144 in^2.)

4. **a.** On a separate sheet of paper, draw a triangle with 40° and 95° angles and a 10.5-cm side between those two angles. Start out by drawing the 10.5-cm side near the bottom of your drawing space.

b. *Calculate* the third angle. It is ________°. Then measure it in your triangle to check.

c. Classify your triangle according to its sides and angles:

It is ______________________________ and ______________________________.

Skills Review 79

1. **a.** On a separate sheet of paper, draw a parallelogram with one 45° angle and sides that are 7 cm and 10 cm long.

   b. Are any of the angles in the parallelogram complementary? If so, which ones?

   c. Are any of the angles in the parallelogram supplementary? If so, which ones?

2. A park is roughly a rectangle that measures 3.3 cm by 1.7 cm on a map with a scale of 1:30,000. What is its approximate area in reality, to the nearest hundred thousand square meters? To the nearest hectare?

3. For each set of lengths below, determine whether the lengths form an acute, right, or obtuse triangle—or *no* triangle. You can construct the triangles using a compass and a ruler or use the Pythagorean theorem.

   a. 7, 5, 10

   b. 12, 15, 9

4. You will now conduct an experiment where the various outcomes are not equally likely to occur. In such a case, we say that the probability model is **not uniform**. Choose from one of the experiments listed or come up with one of your own. Repeat the experiment 100 times and count how many times each outcome occurs. Then calculate the experimental probabilities.

   (1) Place a bunch of socks of two different colors in a bag. Make sure that there are twice as many socks of one color as there are of the other. Take a sock out of the bag, note its color, and then place it back in the bag under the other socks. Then repeat.

   (2) Open a book to a random page. Note whether the page has a picture on it. Then close the book, and repeat the experiment .

   (3) Put three different balls in a bag. Randomly pick one. Put the ball back. Then repeat.

Outcome	Relative Frequency	Experimental probability (%)

Skills Review 80

1. A map has a scale of 1:40,000.

a. Rewrite this scale in the format 1 cm = __________ m.

b. Fill in the table. Give your answers to the nearest tenth of a centimeter.

on the map (cm)	in reality
	350 m
	800 m
	1.4 km
	2.7 km

Write an inequality for the problem, and solve it. Plot the solution set on a number line. Lastly, explain the solution set in words.

2. Shelly wants to buy two blouses that cost $18.99 each and some scarves that cost $7.99 each. How many scarves can she buy if she only has $75 to spend?

3. Choose the proportion that is set up correctly and solve it.

a. $\frac{14 \text{ lb}}{\$91} = \frac{x}{\$143}$

b. $\frac{\$91}{14 \text{ lb}} = \frac{x}{\$143}$

4. One angle is given. Find the measures of the marked angles without measuring.

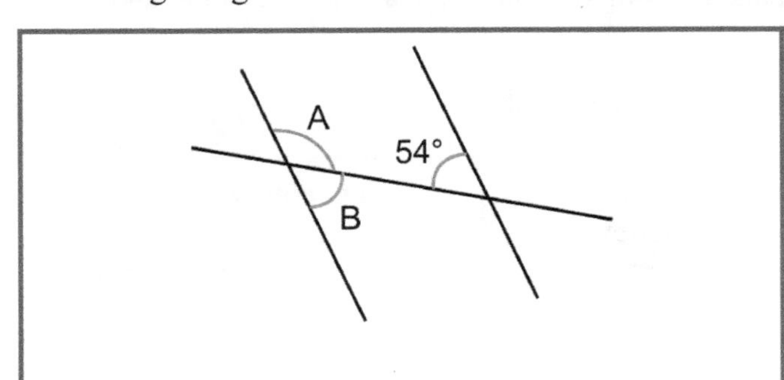

a. ∠A = ______ ° ∠B = ______ °

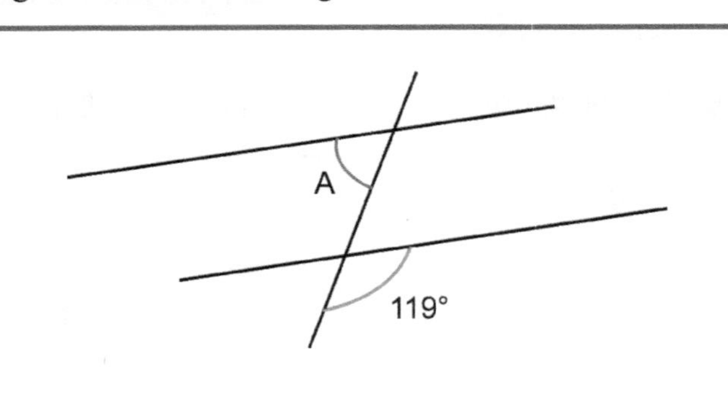

b. ∠A = ______ °

Skills Review 81

1. **a.** Draw a triangle using these three line segments as sides. You will need a protractor.

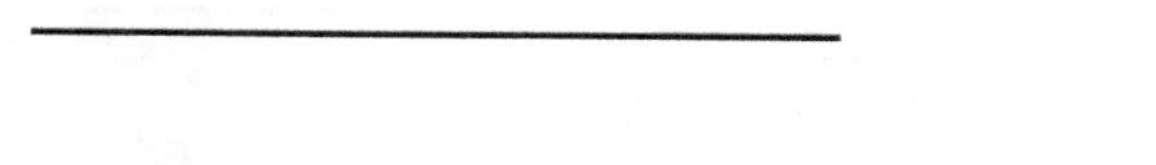

   b. Classify the triangle according to its angles and sides.

2. Caleb took out a loan for $925 for 8 months with an annual interest rate of 11.4%. How much less interest would he have paid if instead he had taken out a loan for 5 months with an annual interest rate of 10.7%?

3. You write each letter of the alphabet on separate pieces of paper, put them all in a bag, and draw one out. Find the probabilities.

   a. P(consonant)

   b. P(the capital form of the letter has no round parts)

4. Calculate the area of an equilateral triangle with 17-cm sides to the nearest square centimeter. Don't forget to draw a sketch.

Skills Review 82

1. Find the value of these expressions to three decimal digits. Use a calculator.
Note: if your calculator doesn't automatically follow the order of operations, you need to use parentheses. Another option is to write the intermediate results down or load them into the calculator's memory.

a. $\sqrt{6.8^2 - 4.3^2}$	**b.** $\sqrt{52.6^2 + 29.17^2}$

2. A bag of dried fruit contains equal amounts of banana chips, pineapple, apricots, and raisins.

a. You take six pieces of dried fruit out of the bag (randomly). Design a simulation for this experiment. Explain your design here:

b. Run your simulation for at least 100 repetitions (but more is better). Record the outcomes.

Hint: If you generate the random numbers at https://www.random.org/integers/, copy the rows of random numbers to a spreadsheet program, such as Excel. You may need to use the "Paste special" command to paste them as "text."

Find the experimental probabilities below.

c. P(none are raisins) =

d. P(exactly 2 are banana chips) =

e. P(2 are pineapple, 1 is a banana chip, and 1 is an apricot) =

(Note: the order does not matter. For example, your set may have the apricot piece first, last, or in the middle.)

Skills Review 83

1. A small cylindrical watering can has an inside diameter of 6.3 inches and a height of 11.8 inches.

   a. Calculate its volume in cubic inches, to the nearest cubic inch.

   b. Convert the volume to gallons, to the hundredth of a gallon.
   (1 gallon = 231 in^3)

2. Fill in the table. Use $\pi \approx 3.14$ and a calculator. Round to one decimal digit.
   Note: do not use rounded answers for continued calculations. If you use an intermediate answer to calculate something else, keep at least four decimal digits for the intermediate answer.

	Circle A	Circle B	Circle C
Circumference	16.0 cm		
Diameter		3.6 in	
Radius			9.1 m

3. Calculate the five-number summary and draw a box plot for the following set of data.

   The results of a high school history test:
   9, 15, 11, 10, 10, 6, 18, 15, 20, 7, 12, 10, 19, 17, 15, 12, 13, 10, 20, 16, 11, 13, 18, 14, 17

Five-number summary
Minimum ________
First quartile ________
Median ________
Third quartile ________
Maximum ________
Interquartile range ________

4. Solve. Round the answers to three decimals.

a. $74 - x^2 = 23$	**b.** $119^2 + s^2 = 15{,}800$

Skills Review 84

1. Find the area in the given units. Then convert it to the other area unit. (1 square mile = 640 acres.)

A rectangle with 0.6 mi and 1.4 mi sides

A = ____________ sq mi

A = ____________ sq mi · ____________ = ____________ acres. (*to the nearest ten acres*)

2. The radius of a certain circle is 9 units. Which expression can you use to calculate the circumference of that circle?

a. $\pi \cdot 9$ **b.** $\frac{9}{\pi}$ **c.** $\pi \cdot 18$ **d.** $\frac{\pi}{9}$

3. The graph shows the average annual income of the employees of a certain business.

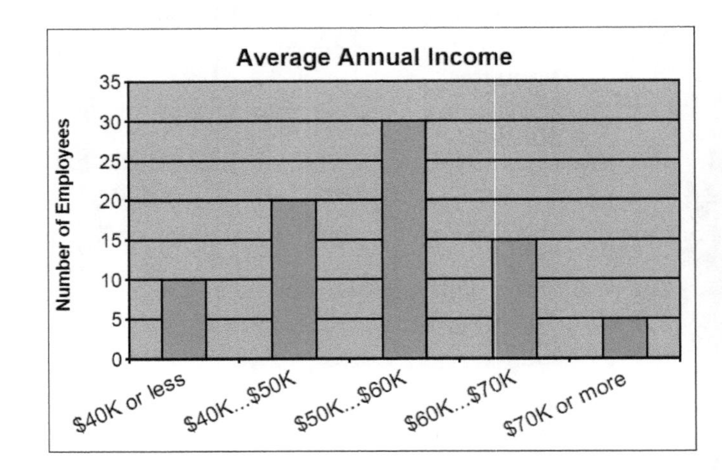

a. If you choose one person randomly, what is the probability they earn between $40K and $50K?

b. If you choose one person randomly, what is the probability they earn at most $70K?

4. You are studying whether the students at a high school prefer spending their free time playing sports, listening to music, shopping, doing volunteer work, or reading.

a. Which of the sampling methods listed below produce a voluntary response sample?

b. Which methods don't give each member of the student population an equal chance to be selected for the sample?

c. Which method will be the most likely to give you a representative (unbiased) sample?

Sampling Methods

(1) You interview 60 students during recess.

(2) You interview the students that you encounter at the local shopping mall on a Saturday.

(3) You interview the first 60 students that leave the school on a certain day.

(4) You choose 60 names randomly from a list of all the students. You call them to interview them.

(5) You post a notice on the school bulletin board, asking students to contact you for an interview. You hope to get at least 60 responses.

Skills Review 85

1. Choose a group of 10 small objects (such as buttons, paperclips, etc.) so that the group has objects of five different colors in it. For example, you could choose 3 blue, 2 green, 1 red, 2 yellow, and 2 pink objects. The experiment will involve choosing an object at random from your group.

   a. Choose an object at random from your group, record the outcome, and put the object back. Repeat this 100 times. Count the frequencies of each outcome. In the table, record the **relative frequencies**—the frequencies written as fractions of the total number of repetitions.

Outcome	Theoretical probability	Relative Frequency	Experimental probability
	_______%	_____/100	_______%

   b. Calculate the theoretical and experimental probabilities for each outcome. Then compare the two: are they fairly close?

   If not, what could have caused the discrepancy?

2. Calculate the volumes to (a) the nearest cubic inch, and (b) to the nearest ten cubic centimeters.

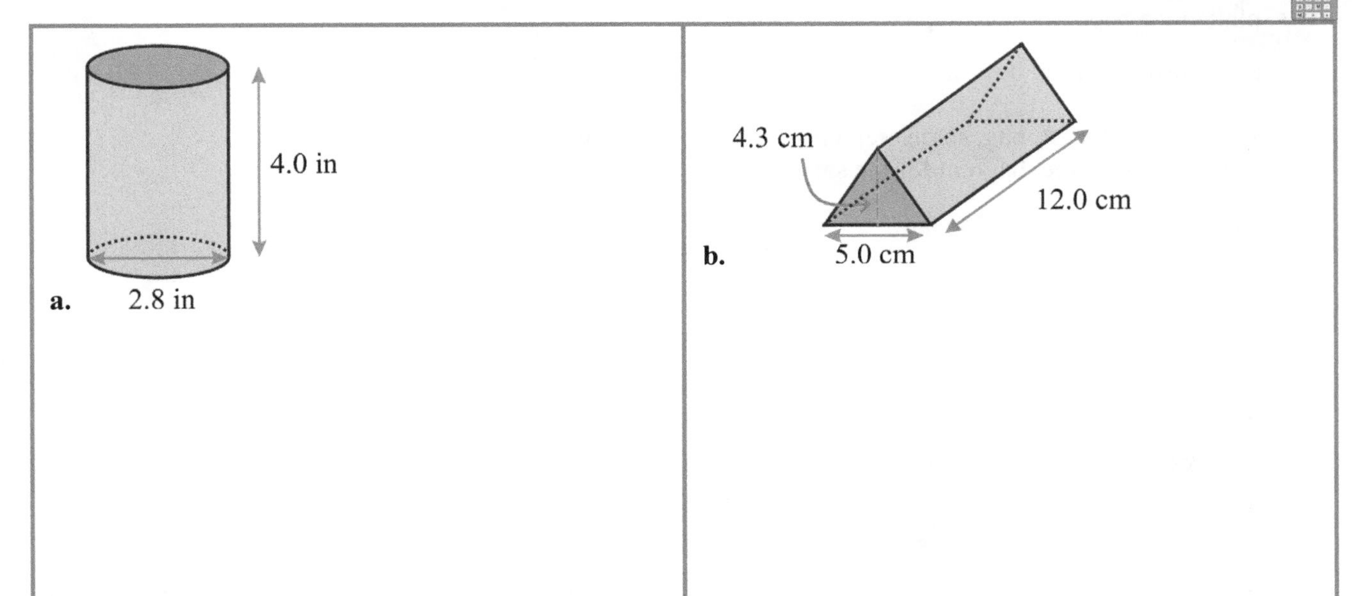

Skills Review 86

1. The sides of a triangle measure 38 mm, 69 mm, and 46 mm. Its altitude is 24 mm. The triangle is enlarged proportionally so that the altitude of the triangle becomes 27 mm. Find the area of the larger triangle to the nearest square millimeter.

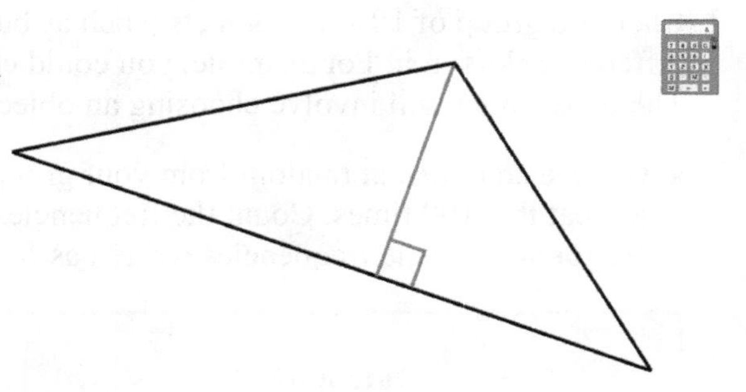

2. A 12-foot ladder is placed against a wall so that the base of the ladder is three feet away from the wall. What is the height of the top of the ladder?

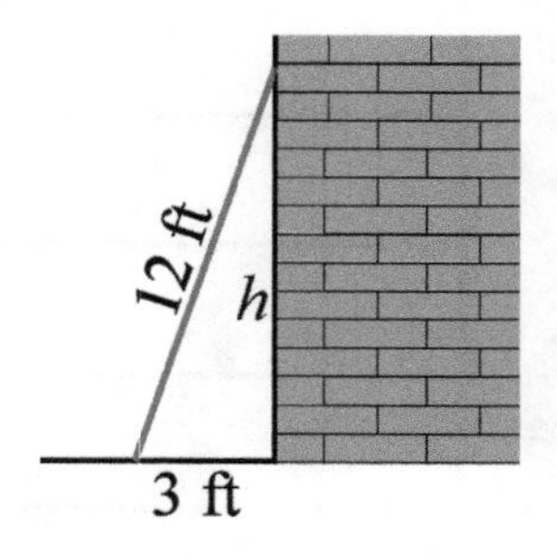

3. Megan is getting a birthday present ready to give to her cousin. She has the following choices:

- a floral gift bag, a sparkly gift bag, or a striped gift bag
- pink crepe paper or blue crepe paper
- a yellow bow, a red bow, a pink bow, or an orange bow

a. Megan chooses a gift bag, some crepe paper, and a bow randomly. On a separate sheet of paper, draw a tree diagram for the sample space.

What is the probability that Megan…

b. … chooses a striped bag, pink crepe paper, and red bow?

c. … chooses blue crepe paper and NOT a striped bag?

d. … doesn't choose a yellow or orange bow?

e. … chooses a sparkly bag and a pink or red bow?

Skills Review 87

1. The data below gives you the weights of chicken eggs laid on a certain day at two different farms.

a. Determine the five-number summaries and draw side-by-side boxplots for the data.

Farm 1 Weights (grams)	
49	52
49	52
49	52
49	52
50	52
50	53
50	53
50	54
50	55
50	55
50	55
51	55
51	56
51	56
52	56

Five-number summary

Minimum: ______

1st quartile: ______

Median: ______

3rd quartile: ______

Maximum: ______

Interquartile range:

____

Farm 2 Weights (grams)	
53	57
54	57
54	57
54	57
54	57
54	57
55	58
55	58
55	58
55	58
55	59
55	59
55	60
56	60
56	60

Five-number summary

Minimum: ______

1st quartile: ______

Median: ______

3rd quartile: ______

Maximum: ______

Interquartile range:

____

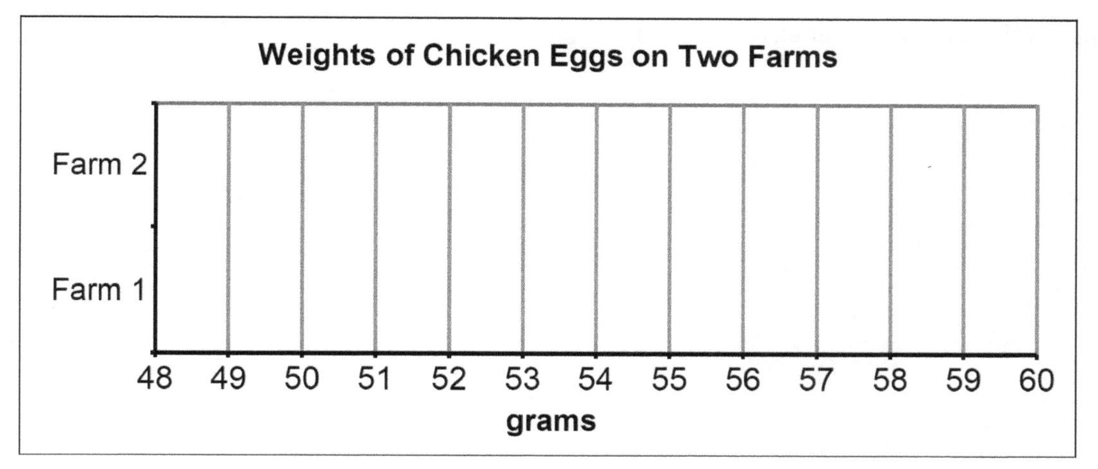

b. Describe the overlap of the distributions.

c. Which farm appears to produce heavier eggs?

d. Is the difference in the medians significant?

Justify your reasoning.

Skills Review 88

1. Researchers studied the social media usage of students from two different colleges. They selected a random sample of students from each college. The double (or back-to-back) stem-and-leaf plot shows the number of hours spent using social media by each member in the sample during one particular month.

Number of Hours Spent Using Social Media in One Month

Group 1 Leaf	Stem	Group 2 Leaf
99988	1	
99998764	2	445899
433322110000	3	033344556677789
53110	4	002488
	5	01

(Note: The stem is the tens digit and the leaf is the ones digit. For example, the first value for Group 1 is 18 and the last one is 45.)

a. Just looking at the two distributions, overall which group appears to spend more hours using social media?

Which group appears to have a greater variability in the number of hours spent using social media?

b. Find the range and median for each sample. The interquartile range is given to you.

Group 1:

Median _________ Range _________

Interquartile range: 6

Group 2:

Median _________ Range _________

Interquartile range: 7

c. Do these values support your answers in (a)?

Skills Review 89

1. Imagine a square with 3-meter sides.

   The area in square meters is $3 \text{ m} \times 3 \text{ m} = 9 \text{ m}^2$.

   The area in square *decimeters* is ______ dm × _______ dm =__________ dm^2.

   The area in square *centimeters* is _________ cm × __________ cm = _____________ cm^2.

   The area in square *millimeters* is ____________ mm × _____________ mm = ___________________ mm^2.

   So, 9 m^2 = _________ dm^2 = ____________ cm^2 = _______________ mm^2.

2. The bottom of a pyramid is a square with 8.0-cm sides, and the height of each triangular face of the pyramid is 6.5 cm.

   a. Sketch a net for the pyramid.

   b. Calculate the surface area of the pyramid.

3. The owner of a natural skin care products company wanted to know which massage oils are most popular among massage therapists. She randomly chose participants of two different massage therapy seminars to interview, asking them what their favorite oil was. The results are in the table at the right.

   What conclusions can you draw from the data?

Oils	**Seminar 1 (Sample 1)**	**Seminar 2 (Sample 2)**
Almond Lavender	10	8
Coconut	3	5
Jojoba Rose	5	3
Sweet Almond	7	9
Totals	25	25

Skills Review 90

1. The chance that a random employee at a certain office arrives at work early is 30%. One day, the office supervisor chooses eight of the employees at random. Design a simulation to study this situation.

   Explain your design:

   Now run the simulation. Repeat the experiment at least 100 times, but more is better. Record each outcome. Then count how many of your outcomes represent zero employees arriving at work early, one employee arriving at work early, and so on, and calculate the corresponding probabilities.

Results of simulation		
Employees who arrived at work early	**Relative Frequency**	**Experimental Probability**
0		
1		
2		
3		
4		
5		
6		
7		
8		
TOTALS		**100%**

Now use the results of the simulation to answer the following questions:

a. What is the probability that two of the eight employees have arrived at work early?

b. What is the probability that only one employee has arrived at work early?

c. What is the probability that none of the eight arrived at work early?

d. What is the probability that *at most* 2 of them have arrived at work early?
Hint: add the probabilities from (a), (b), and (c).

e. What is the probability that *at least* 3 of them have arrived at work early?

CPSIA information can be obtained
at www.ICGtesting.com
Printed in the USA
LVHW021719160921
697984LV00004B/16